MILITARY LEADERSHIP IN THE 21ST CENTURY

Science and Practice

Military Leadership in the 21st Century

Science and Practice

Kim-Yin Chan
Star Soh
Regena Ramaya

CENGAGE Learning™

Andover • Melbourne • Mexico City • Stamford, CT • Toronto • Hong Kong • New Delhi • Seoul • Singapore • Tokyo

CENGAGE Learning

Military Leadership in the 21st Century: Science and Practice
Kim-Yin Chan, Star Soh and Regena Ramaya

Publishing Director:
Paul K. H. Tan

Senior Publishing Executive:
Lian Siew Han

Assistant Publishing Manager:
Pauline Lim

Production Executive:
Cindy Chai

Senior Product Director:
Janet Lim

Senior Product Manager:
Charles Ho

Copy Editor:
T. C. Seow

Cover Designer:
Lee Meng Hui

Cover Images:
(Top) Lalage Snow/Corbis, (bottom) US Navy/Cynthia Griggs/Corbis

Compositor:
International Typesetters Pte Ltd

ISBN-13: 978-981-4352-64-2
ISBN-10: 981-4352-64-0

Cengage Learning Asia Pte Ltd
5 Shenton Way #01-01
UIC Building
Singapore 068808

Cengage Learning is a leading provider of customized learning solutions with office locations around the globe, including Andover, Melbourne, Mexico City, Stamford (CT), Toronto, Hong Kong, New Delhi, Seoul, Singapore and Tokyo. Locate your local office at **www.cengage.com/global**

Cengage Learning products are represented in Canada by Nelson Education, Ltd.

For product information, visit our website at **www.cengageasia.com**

Printed in Singapore
1 2 3 4 5 14 13 12 11

Contents

List of Figures xi
List of Boxes xiii
List of Tables xv
List of Abbreviations xvii
About the Authors xix
Foreword xxi
Preface xxiii
Acknowledgements xxvii

Chapter 1 The Science and Practice of Military Leadership **1**
Overview 1
Understanding 'Leadership' 2
Military & Non-Military Definitions of Leadership
Why Leadership Is Not Well Understood
'Military' Leadership 8
Doctrinal (Military) versus Academic (General) Definitions
Context-free versus Context-Specific Knowledge
Leadership versus Management
Leadership versus Command
The Practice of Military Leadership Today 14
Why Military Leadership Matters
What Shapes Doctrines and the Practice of Military Leadership?
The Study of Military Leadership 17
Different Approaches
Why Study the Science of Military Leadership
Summary 22

Chapter 2 Challenges and Stress in Military Operations 27
Overview 27
Challenges Facing Commanders Today 28
The Challenge of Multiple and Complex Security Threats
The Challenge of Ill-Defined Enemy
The Challenge of Emerging Technologies
The Challenge of Civilian Involvement
The Challenge of Media Everywhere
Conditions and Stressors in Operations and Battlefield 33
Harsh Physical Environment and Working Conditions
Deprivation and Fatigue
Enemy Threat
Fog of War
Destruction and Casualties
The Stress Process and Outcomes 36
A Model of Stress
The Stress Process
Stress Response: Fight, Freeze or Flight (and More . . .)
Combat Stress Reaction and Casualties
Stress Management and Building Resilience 39
Stress Management Techniques
Hardiness and Battlemind
Combat Stress Reaction Management
Traumatic Event Management
Behavioral Science and Mental Health Professionals
Summary 44

Chapter 3 Soldier Performance: The Psychology of Human Behavior in Combat 49
Overview 49
Soldiers' Performance 50
Task Performance
Contextual Performance
Counterproductive Work Behavior
A Framework of Soldiers' Performance
The Psychology of Human Behavior 54
Behaviors and Its Determinants

Situational Factors 55

The Power of Authoritative Figures to Elicit Compliance

Group (Informational) Influence on Individuals to Conform

Group/Unit (Normative) Influence on Individuals to Conform

Role Influence

Crowd Influence: Deindividuation Effect

Crowd (Bystander) Influence: Diffusion of Responsibility Effect

Weak and Strong Situations

Fundamental Attribution Error

Person Variables 63

Knowledge and Skills

Abilities and Competencies

Beliefs and Values

Attitudes

Personality

Motivation

Summary 67

Chapter 4 Understanding Soldier Motivation and Morale 71

Overview 71

What is Motivation and Types of Motivation 72

Definition and Nature of Motivation

Underlying Causes

Theories of Motivation 73

Maslow's Hierarchy of Needs Theory: Our Needs Drive Our Behavior

Herzberg Two-Factor Theory: Different Factors for Satisfaction and Dissatisfaction

Reinforcement Theory: Motivation via Carrot and Stick

Expectancy Theory: Motivation via Beliefs and Expectations

Equity Theory: Motivation via Fairness Perceptions

Self-Determination Theory: Moving from Extrinsic to Intrinsic Motivation

Combat Motivation — Why Soldiers Fight 79

Long History

What is Combat Motivation? What Are the Underlying Factors?

Basic Biological and Physiological Needs

Cohesion

Leadership
Self-Confidence
Ideology
Espirit de Corps
Morale 85
Commonly Used but Lacks a Common Definition
Importance and Outcomes of Morale
Factors Influencing Morale
Summary 89

Chapter 5 Military Leaders: Born or Made? 93
Overview 93
Early Trait and Behavioral Approaches: 1900s–1970s 94
The Early Search for 'Leadership Qualities'
The Early Search for Leadership Behaviors
Revival of Trait Approach: 1970s to Present 96
Realizing that 'Leadership Is not so Simple'
Enhanced Understanding of 'Traits'
Understanding Why and How Traits Relate to Leadership
What We Know about Leadership Behaviors Today 102
Task, Relations and Change Behaviors
'Cognitive Skills' Needed for Military Leadership
'Competencies' Required for Military Leadership
Applying the Science of Leadership Traits and Behaviors 105
Traits and Military Leader Selection
Military Leadership Development
Summary 110

Chapter 6 Military Leadership: A Matter of Styles or Values? 115
Overview 115
Early 'Styles Approach: 1940s–1970s 116
Origins of 'Styles' Approach
What Are Leadership Styles?
The 'New Paradigm' for Leadership Styles 120
From Transactional to Transformational Leadership
Burns on 'Transforming Leadership'
Bass' Model of Transformational Leadership
Military Research Supporting Bass' Theory

What We Now Know About 'Styles' ... 126
Military Forces on 'Leadership Styles' ... 127
'Styles' in U.S. Army Doctrine
Other Military Forces on Styles
Values and Military Leadership ... 129
Why Values Matter
Perspectives on Values and Military Leadership
Beyond Styles and Values: 'Authentic Leadership' ... 132
Summary ... 134

Chapter 7 Science and Practice of Leading Military Teams ... 139
Overview ... 139
The Science of Leading Teams ... 140
Evolution of Team Research
Basic Terminology
Models of Team Effectiveness
Command Team Effectiveness Framework (CTEF)
Leadership Factor in Team Effectiveness Models
Team leadership — The 'Functional Approach'
Leader Behaviors in Teams
Trust in Teams and Trust in Leadership
The 'Practice' of Leading Teams ... 153
Team Building — What Is It?
Team Development — What Must a Leader Do?
Summary ... 160

Chapter 8 Leading Military Teams for Adaptive Performance in Complex Environments ... 167
Overview ... 167
Team Performance Challenges in Complex Settings ... 168
Cognitive, Behavioral and Affective Demands
Implications for Team Leaders
The Team Training Approach ... 171
Key Competencies for Teams in Complex Environments ... 173
Adaptability in Teams
Critical Thinking Skills for Team Decision Making
Shared Mental Models (SMMs)

Strategies for Training Adaptive Teams 180

Team Adaptation and Coordination Training (TACT)

Team Dimensional Training (TDT) and Guided Team Self-Correction Technique

Stress Exposure Training (SET)

Cross Training

Leader Actions for Adaptive Team Performance 187

Leader Behaviors that Support Team Learning

Leader Roles that Develop Adaptive Expertise in Teams

Leader Actions that Develop Critical Thinking

Leader Actions that Develop SMM in Teams

Leader Actions for Team Self-Diagnosis and Self-Correction

Summary 192

Chapter 9 Leading in Hierarchical, Learning Military Organizations 197

Overview 197

Understanding Organizations 198

Organizational Structure

Organizational Culture

The Nature of Leadership in Organizations

Levels of Military Leadership 202

'Levels' in U.S. Army Doctrine

'Levels' Based on Stratified Systems Theory

'Levels' in the Canadian Forces

Do New Forms of Military Organization Need More or Less Leadership? 206

Learning Military Organizations 208

Peter Senge on 'Learning Organization'

Organizational Development, Change and Learning

'Demystifying' the Learning Organization

Case Study: U.S. Army's Transformation to a Learning Organization

Military Forces on Leadership and Organizational Change 215

Summary 216

Chapter 10 Leading in the 21st Century Profession of Arms 221

Overview 221

What is a 'Profession' 222

A Brief History of 'Professions'

Ways to Think About 'Professions'

The Military as a 'Profession' 227

Why Did Military Forces 'Professionalize'?
Who Is a Military Professional?
Unique Features of the Military Profession
21st Century Threats to the Military Profession 231
Organizational-Level 'Threats' to the Military Professional Institution
Individual-Level 'Threats' to the Professional Military Career
How Rising 'Boundarylessness' Affects the 21st Century Profession of Arms
Strategies to Revitalize and Strengthen the Profession of Arms 238
Strengthening Professional Ethos and Motivation
Strengthening Professional Organization
Leading in the 21st Century Profession of Arms: Emerging Themes 244
Summary 246

Index 251

List of Figures

1-1	Leadership is about influencing people in situations to accomplish outcomes and purposes in a certain context	4
1-2	Singapore military's "Leadership 24-7" framework	9
1-3	Leadership, command and management in the military context	14
2-1	Range of U.S. Joint Operations	29
2-2	A model of stress in the military	36
3-1	A Framework of Soldier Performance	53
4-1	Maslow's Hierarchy of Needs	73
4-2	Herzberg's Two-Factor Theory	74
4-3	Vroom's Expectancy Theory	76
4-4	Equity Theory	77
5-1	The language of "traits" today	98
5-2	Chan & Drasgow's (2001) theoretical framework linking traits with various leadership outcomes	101
5-3	Example of two military leadership competency models	105
6-1	Tannenbaum and Schmidt's one-dimensional "continuum" versus Blake and Mouton's two-dimensional "grid" representation of the relationship between "task" and "people" leadership orientations	117
6-2	Hersey & Blanchard's (1969) Situational Leadership Theory suggests "if-then" rules for various styles depending on the followers' "maturity" or "developmental level"	118
6-3	Main ideas in Fielder's Contingency Theory of Leadership	119
6-4	Main ideas and findings related to Bass' (1995, 1998) theory of leadership	123
6-5	Canadian Force's Spectrum of Leadership Styles	128
7-1	A representation of the I-P-O model of team effectiveness	143
7-2	The CTEF Model	146
7-3	A framework for military team building	155

7-4 A model of systematic team building in the military 156
8-1 Team Demands in the 21st century 170
8-2 Knowledge and skills for leading military teams in complex environments 171
8-3 Types of Team Interventions 172
8-4 Critical Competencies for Adaptive Team Performance in Complex Environments 173
8-5 The STEP process 178
8-6 Team Training Strategies 180
8-7 Team Dimensional Training Framework 183
9-1 Single and Double-loop learning 211
10-1 Prototypical features of "professions" 226
10-2 Relationship between individual military professionals with the "profession of arms" (organization or institution) in each state 232
10-3 21st century threats to the military profession 233
10-4 Strategies by military forces and nations to strengthen military professionalism 238

List of Boxes

1-1	"Leadership" in various Military Doctrines	3
1-2	"Leadership" in Academic Textbooks	4
2-1	Mental Health of U.S. soldiers in Operation Iraqi Freedom (OIF) in 2009	39
2-2	Battlemind	42
3-1	The My Lai Massacre of Civilians during the Vietnam War in 1968	56
3-2	The Abu Ghraib prisoners abuse in Operations Iraqi Freedom	61
4-1	How to assess Morale and use the findings?	88
8-1	USS Vincennes Incident and the TADMUS program	169
8-2	Three phases of SET	185
8-3	Leader Actions in the Team Learning Cycle	189
10-1	"Reprofessionalization" of Canadian Forces	239

List of Tables

1-1	Leaders versus Managers	11
2-1	Minimum Operational Sleep Requirement	34
5-1	The Five Factor Model (FFM) of personality	99
5-2	Task, relationship and change-oriented leadership behaviors	102
9-1	Henri Fayol's 14 Principles of Organization	199
9-2	U.S. Army's three levels of leadership with Jacques' seven levels	205
9-3	Senge's Five Disciplines of a Learning Organization	209

List of Abbreviations

AAR	After Action Review
ADF	Australian Defence Force
BICEPS	Brevity, Immediacy, Contact, Expectancy, Proximity & Simplicity
BRM	Bridge Resource Management
CDM	Classical Decision Making theory
CF	Canadian Forces
CIC	Combat Information Center
CRM	Crew Resource Management
CSR	Combat Stress Reaction
CTA	Cognitive Task Analysis
CTEF	Command Team Effectiveness Framework
CWB	Counterproductive Work Behavior
E-P	Expectancy-Performance probability
FFM	Fire Factor model of Personality
FM	Field Manual
IAMPS	International Applied Military Psychology Symposium
IDF	Israeli Defense Forces
IMTA	International Military Testing Association
IPK	Interpositional Knowledge
I-P-O	Input-Process-Outcome approach
IUS	Inter-University Seminar on Armed Forces & Society
LLAAD	Leader-led After Action Debriefing
MOOTW	Military Operations Other Than War
MLQ	Multifactor Leadership Questionnaire
MTL	Motivation to Lead construct
N.S.	National Service

NATO	North Atlantic Treaty Organization
NATO RTO	NATO Research & Technology Organization
NDM	Naturalistic Decision Making theory
NTU	Nanyang Technological University
OD	Organization Development
OE	Organizational Effectiveness
OIF	Operational Iraqi Freedom
OLM	Organizational learning mechanism
ONR	U.S. Office of Naval Research
OSS	Office of Strategic Services
PIE	Proximity, Immediacy, Expectancy
P-O	Performance-Outcome probability
PSC	Private security contractor
PTSD	Post-Traumatic Stress Disorder
R/M	Recognition/Meta-cognition model
SAF	Singapore Armed Forces
SET	Stress Exposure Training strategy
SMM	Shared mental model
STEP	Story-Test-Evaluate-Plan process
SWOS	Surface Warfare Officers School
SWOT	Strength-Weaknesses-Opportunities-Threats
TACT	Team Adaptation & Coordination Training
TADMUS	Tactical Decision Making Under Stress
TDT	Team Dimensional Training
TRC	Task, Relationship & Change behaviors
VIE	Valence-Instrumentality-Expectancy theory

About the Authors

Kim-Yin CHAN

Dr Kim-Yin Chan joined Nanyang Business School as an Associate Professor in 2009. Before that, he served in the Singapore Armed Forces as an Infantry Officer in the SAF for almost 25 years and held the rank of Lieutenant-Colonel. His last two appointments in the SAF were Head of Leadership Doctrine & Research and Head of the Continuing Education (CE) Project Office in SAFTI Military Institute from 2002-2009. He received his Bachelor degree in Psychology with First Class Honors from the University of London in 1988 and his Masters and Ph.D. degrees in Industrial-Organizational Psychology from the University of Illinois at Urbana-Champaign in 1997 and 1999 respectively.

Star SOH

Dr. Star Soh joined the Nanyang Business School as an Associate Professor in 2008. His previous appointments were Chief Psychologist of the Ministry of Defence (MINDEF) and Head of the Applied Behavioral Sciences Department. He retired as a Lieutenant-Colonel after 25 years of service. Dr. Soh Star obtained his Bachelor degree in Psychology with First Class Honors from University of Manchester, U.K., Postgraduate Diploma in Personnel Management from Singapore Institute of Management, and Masters and Ph.D. degrees in Industrial-Organizational Psychology from the Ohio State University, U.S.A.

Regena RAMAYA

Ms. Regena Ramaya is currently the Director and Principal Consultant of OPAS Consulting. She is an Occupational Psychologist by profession, and retired from the Singapore Armed Forces at the rank of Major in 2008. In her 25-year military career, she held staff, instructional and command appointments. Her last appointment was Head of Leadership Assessment Systems in the SAF Centre of Leadership Development (SAF CLD). She has an Honors degree in Psychology from the University of Manitoba (Canada) and a Masters degree in Occupational and Organizational Psychology from the University of Surrey (U.K.).

Foreword

It is with great pleasure that I introduce *Military Leadership in the 21st century: Science & Practice.* This book is written by three recently-retired Singapore military officers who have been responsible for designing and conducting a new undergraduate course to support the undergraduate-level professional military education of Singapore Armed Forces (SAF) regular officers. This book, and the course which it supports, is unique in that it aims to provide junior officers with an introductory-level appreciation of the key concepts and issues related to military leadership from a social-behavioral science perspective, which is the basis for many military leadership doctrines today. Another unique feature of the book is its emphasis on the military as the "Profession of Arms", which is an important distinguishing factor for leading in the military.

Until the 1980s, the Singapore Armed Forces did not have its own leadership doctrine. In its early years, leadership in the SAF was learned primarily through experience and observation, with a sprinkling of catch-phrases such as "lead by example" or "duty, honor, country" that made intuitive sense and were seldom challenged or questioned. To ensure that leadership or its development would not be left to chance, the behavioral scientists in the SAF were tasked in the 1990s to develop doctrinal frameworks and tools to support systematic leadership development throughout the organization. In 2002, the SAF established a Center for Leadership Development with an enhanced leadership framework and competency model so that tools, practices and positive habits could be proliferated in both in the training establishments and in the workplace. Throughout their military careers, all three authors were influential in helping the SAF to ensure that its leadership development system is based on sound, scientific principles. Given their backgrounds as behavioral scientists with military leadership experience, I couldn't think of a more qualified group of writers for this book.

When the SAF-NTU Academy was established in 2009 as part of a strategic partnership between Nanyang Technological University (NTU) and the SAF to drive professional military education in Singapore, I was happy to learn that this group of former SAF officers agreed to design and conduct the new military leadership course. I was even more delighted when they decided to write this book to codify their highly specialized knowledge on a complex and evolving subject like military leadership for future generations of military officers.

In my view, leadership is both an art and a science that is ultimately developed through constant study and practice. Leadership is an art to the extent that it involves actions and decisions that are unique and to each leader and his or her circumstances, and as an art, "leadership" is subject to personal interpretation and philosophy. This book however emphasizes the science of military leadership — much of which is funded by military forces that have an interest in systematically selecting and developing leaders, and which are interested in the effective practice of leadership in what is a "high stakes" business where lives are at stake. Based on this research, leadership can be studied as a science because there are very useful frameworks and concepts, anchored on years of research and studies, which leaders can draw from for their understanding and practice of leadership.

Twenty-first century military operations are highly complex, dynamic, ambiguous and stressful. In such environments, there are no fixed clear-cut solutions to leadership challenges and leaders need to constantly adapt to changing demands and circumstances. The authors emphasize in Chapter 1 that ultimately, leadership is best learned through an experiential learning process that includes actual experience, observation, and reflection to relate these to established principles and concepts, and active experimentation to "test" their new learning. I fully agree with this approach and would encourage all future military officers to use the concepts and principles in this book for reflecting on their leadership — so that they may improve their ability to influence their soldiers in the accomplishment of their missions!

Brigadier General (N.S.) Jimmy Tan Cheng Yaw
Director, SAF-NTU Academy

Preface

Bring up the subject of military leadership and most will think of the study of great military Generals and warriors — especially of the past wars and battles — such as Rommel, Patton, Montgomery, Slim, Jackson, Lee, Genghis Khan, etc. Yet, take a close look at the doctrinal manuals, news reports and professional writings on leadership in today's military forces and one notices that the focus is on leadership expected of all personnel in the organization — from the "strategic privates" to the generals, and from uniformed to non-uniformed personnel — all of whom are expected to influence others in their daily lives and in preparation for a wide spectrum of future military operations.

We wrote this textbook with the belief that the concept of military leadership — like the study of leadership itself — has shifted over the last century from a "Great Man" perspective to one that focuses on the idea of "leadership as influence". Military forces promulgate doctrines to guide practice among their personnel. A quick look at many of the military doctrines on leadership shows a convergence with concepts and ideas found in the social scientific literature on the subject. Today, this convergence is so dramatic that one even finds some military forces like the Canadians introducing academic concepts like "distributed leadership" in their doctrines to shape its practice. We believe that many military leadership doctrines have as their foundations principles from social-scientific research, which is in turn often driven by the contemporary operational and organizational military challenges, and supported by military funding or samples. This shift, in turn, reflects the growing influence of the social-behavioral sciences in the military, alongside the "hard", physical and engineering sciences.

The authors of this book were military officers who had the privilege of gaining and applying their knowledge of the social-behavioral sciences in the military context. In this book, we use the term "social-behavioral sciences" broadly to include psychology (including, its many sub-fields such as social, clinical, individual differences, industrial-organizational psychology, etc.), sociology, organization and management science. Today, much of this social-behavioral science knowledge of leadership is published and shared within highly specialized communities such as defense research centers, doctrine departments, military academies, seminars, conferences and academic journals. This community comprises both military and academic researchers and practitioners who have an interest in the many aspects of military leadership — from the selection and classification of military leaders to their professional socialization, training and education, from their performance on and off the field, to their continuous leadership development. This community is itself highly disparate — given the highly specialized and specific interests that each sub-group possesses. For example, those who focus on military leadership selection may not be familiar with the work done by those who focus on the leaders' role in preventing and managing soldiers' stress in military operations.

Our motivation for writing this book is to bring to focus the vast knowledge that we have acquired from the many perspectives on this subject and to present it in a useful way to educate and inform military leaders. In our careers as military behavioral scientists, we have always been fascinated by the symbiotic relationship between the military and the growth in certain knowledge domains of psychology — one of which is knowledge related to the subject of leadership — from the selection and classification of leaders, to the study of leadership behaviors and styles, human performance and decision making under stress, leadership and organizational climate, and even more recently, the study of team leadership. How did this relationship between the military and the social-behavioral sciences come about? We believe the key idea lies in the systematic way in which both the scientific and military communities address issues and challenges in the practice of leadership. Hence, we subtitled this book: "Science & practice".

The book aims to provide junior commanders or entry-level military officers with an introductory-level appreciation of the key concepts and issues related to military leadership from a social-behavioral science perspective, and, as it is commonly defined in the doctrines of modern military forces. The current edition of this book focuses on "direct-level" or "team leadership" and has three main themes: (1) the human dimension of military operations and organization; (2) the main approaches to direct or team-level leadership, including the trait-approach, values-based or ethical leadership, leadership styles and behaviors; and, (3) the challenges of leadership in 21st century military operations, organizations, and, the 21st century military profession. Future editions may expand to include chapters on organizational and strategic level leadership.

The book begins with an overview of the concept of military leadership and its articulation in modern military doctrines which guide its practice, and its basis in the social-behavioral sciences. The next three chapters provide a basic introduction to the context of military leadership in modern operations, including the task/environmental factors and the human factor (performance and motivation). Chapters 5 and 6 provide an overview of leadership theory and research, with focus on the trait, behavioral, style and values approaches. An attempt is made to show the relevance and application of this knowledge in modern military forces. The next 2 chapters focus on the team-level context of military leadership. Chapter 7 provides the basic principles of leading teams, while Chapter 8 surveys the unique challenges of team functioning in high-stress, high-tempo and high-technology military operations. Chapter 9 extends the discussion to the organizational context — emphasis is given to the nature and challenges of leading in transforming and learning military organizations (as opposed to leading in traditional military hierarchies). The concluding chapter discusses the challenges of military leadership in a changing profession of arms.

In each chapter, we attempt to relate the social scientific knowledge to various modern military doctrines on leadership. This book therefore draws on primary social-behavioral science research reports and writing, and various military doctrinal documents that are publicly available in the English language (e.g., Australia, Canada, Israel, New Zealand, Singapore, U.S., U.K.). Besides introducing key leadership-related concepts and issues, readers are also introduced to key events, institutions and social-behavioral scientific research programs that have shaped contemporary military leadership doctrines and systems (e.g., leadership selection, leadership & team training and development, organizational learning & change, etc.). The ideas are presented in a general social-scientific style as opposed to a purely military "doctrinal", or a historical/biographical style. Effort is therefore made to highlight the theories, concepts and findings from military-related research programs, and the military interests in these programs (e.g., "why" the military decided to fund or sponsor these studies; or how these issues are relevant to the military).

In our view, modern military forces — which are ultimately concerned with the survival and effectiveness of their soldiers and civilians on the battlefields of today and tomorrow, and that are constantly concerned with integrating human and technical systems — have been strong users and supporters of the social-behavioral research. On the other hand, the social-sciences have also benefited from the military forces that have given them access to large research samples and more importantly, practical research questions. This symbiotic relationship has not only led to significant advances in the social-behavioral sciences, but also enabled military forces to be at the forefront of human capital management, development, health and welfare. Our intent in taking the approach in this book is to "bring back to the military" the knowledge that it has generated through its funding and sponsorship of military leadership research.

We hope that military leaders may apply the principles from this science-based knowledge of leadership as they go about influencing people in the course of their work, and, as they establish policies that affect the lives of those that they lead. We therefore dedicate this book to all military leaders and the scientists and practitioners — past, present and future — who have and will continue to build this knowledge domain to the benefit of both the military profession and the social-behavioral sciences.

Kim-Yin Chan, Star Soh & Regena Ramaya
June 2011

Acknowledgements

Our impetus for writing this book is the new course, Military Leadership, which we were entrusted to design and conduct as part of the undergraduate professional military education program for all junior Singapore military officers. We thank the Singapore Armed Forces (SAF) and the Nanyang Technological University (NTU) for giving us the opportunity to conceptualize, plan and design the new course. In particular, we would like to express our appreciation to the Director of the SAF-NTU Academy, Brigadier General (N.S.) Jimmy Tan, and his staff for their kind assistance, encouragement and support for our efforts throughout the development of the course and for the production of this book. We also thank Colonel Sukh Singh and Lt. Colonel Adrian Chan from the Singapore Armed Forces Center of Leadership Development and Professor Ang Soon from the Nanyang Business School for supporting the design and conduct of our new Military Leadership course.

We would like to acknowledge the key persons who supported our career development as military behavioral scientists, in particular K.C. Ong and Tan Kee Yong, who both believed in the importance of the applied, military behavioral sciences and who encouraged us to pursue our further education in psychology. We would also like to thank Goh Chee Kong for mentoring us as military officers, and Micha Popper for coaching and training us as applied military behavioral scientists. We acknowledge the key role played by our former Chief of Defence, Lt. General (N.S.) Lim Chuan Poh in initiating and driving the Singapore military's systematic approach to leadership development after the turn of the new millennium.

The effort to research and write this book was a two-year journey during which we received much encouragement, advice and feedback from our friends and colleagues. In particular, we thank Lim Kwee Hoon, Tay Li Ping, Allister MacIntyre, Stefan Seiler, Jamie Cullens and Paul Bartone, and also our students from the first run of the Military Leadership course in 2010. We are grateful to our publisher, Cengage Learning, for believing in the potential of our book project. We especially thank Han Lian Siew who worked closely with us throughout the editorial and production process to ensure the best outcome possible. We also acknowledge the kind support of the Director of

Public Affairs, Colonel Desmond Tan, and his staff at CyberPioneer, for granting us permission to use their excellent photographs.

Finally, we would like to thank our beloved families for their patience, encouragement, and loving support over the past two years as we worked on this book.

1

The Science and Practice of Military Leadership

Source: CyberPioneer, Ministry of Defence, Singapore

OVERVIEW

Ask any man in the street what they associate with "military leadership" and many would describe the stereotype of rigid, authoritarian, and sometimes "heroic", "great" leaders or "generals". Yet, if one examines the leadership doctrines of military forces today, one is unlikely to find such descriptions of the concept. Instead, one will find in modern military doctrines definitions, principles and concepts that are in fact very similar to those found in academic textbooks on leadership. Look harder, and one will even notice references made to social-behavioral scientific research on leadership, of which a large number are funded by the military or conducted with military samples.

This chapter provides an introduction to the contemporary, social-scientific study of military leadership, which is the basis for many modern military leadership doctrines that guide its practice. First, similarities in how leadership is defined in both military doctrines and in academic textbooks are observed. The chapter then proceeds to discuss some of the general challenges of understanding leadership, and the nature of "military" leadership. An observation is made that the contemporary understanding of leadership in both the academic and military worlds today has roots in the debate on the difference between leaders and managers in the 1970s and 80s; and, how military forces further attempted to clarify the meaning and practice of leadership by contrasting it with the concept of "command" in the 1990s. Next, reasons "why military leadership matters" and the factors that shape the thinking on leadership in modern military forces are discussed. The chapter concludes by discussing different approaches to the study of military leadership, and makes a case for a scientific approach that is often the basis for military leadership doctrines today.

UNDERSTANDING "LEADERSHIP"

Military & Non-military Definitions of Leadership

The first decade of the 21st century witnessed a rapid increase in the attention given by various military forces around the world to leadership and leadership development. From 2001 to 2005, the military forces of various countries such as the U.K., Canada, Australia and Singapore established leadership centers (see Horn & MacIntyre, 2006), and introduced their military leadership frameworks, models or doctrines. In 2006, the U.S. Army's well-established "Be-Know-Do" leadership doctrine was also revised, and a new field manual (FM6–22) was issued in 2006 to replace the 1999 field manual (FM22–100).

Why do military forces continue to pay attention to the concept of leadership in their doctrines? How do modern military forces define the concept of leadership in their doctrines today? What is the basis for their doctrinal thinking on leadership? How is the concept of military leadership similar or different from how it was conceptualized in the past or in non-military, academic writings? Military forces promulgate doctrines to guide practice. Box 1-1 summarizes the leadership doctrines of various military forces today while Box 1-2 provides the definitions of leadership from some of the most well-established academic textbooks of leadership. If one were to compare the definitions in Boxes 1-1 and 1-2, one quickly notices that the contemporary military definitions of leadership are very similar to the definitions adopted in general texts in that all describe leadership as a *process* of influence. In this approach, the leadership is not only about the leaders or the people who hold leadership positions. Rather, leadership refers to the *process* of influence which involves the leaders and their followers in a certain situation trying to achieve objectives or outcomes and purposes in a certain context (including mission or task, organizational or cultural, time pressure or stress, etc.). This is illustrated in Figure 1-1.

BOX 1-1

"Leadership" in various Military Doctrines

- **Australian Defence Force:** Leadership is defined as "the process of influencing others in order to gain their willing consent in the ethical pursuit of missions" (p. 1–5, Australian Defence Headquarters, 2007). This definition attempts to capture six "components" that are thought by the Australian military to be required for a process to be called a leadership relationship. These are: Influence (or motivating or inspiring others), followers (who have a choice to follow or not), goals (or vision or mission), purposeful action (by the leader; as opposed to the nature of the outcome or result of the action), consent (of the followers), and the ethical dimension (or morality of leadership).
- **U.S. Army:** "Leadership is the process of influencing people by providing purpose, direction, and motivation while operating to accomplish the mission and improving the organization" (p. 1–2, U.S. Army, 2006). This is contrasted with a definition of the "Army leader", who is "anyone who by virtue of assumed role or assigned responsibility inspires and influences people to accomplish organizational goals. Army leaders motivate people both inside and outside the chain of command to pursue actions, focus thinking, and shape decisions for the greater good of the organization" (p. 1–1, U.S. Army, 2006).
- **Canadian Forces (CF):** "Leadership may be defined as directly or indirectly influencing others, by means of formal authority or personal attributes, to act in accordance with one's intent or a shared purpose" (p. 3, Canadian National Defense, 2005). This definition is presented together with a definition of "effective CF leadership" which is defined as "directing, motivating, and enabling others to accomplish the mission professionally and ethically, while developing or improving capabilities that contribute to mission success" (p. 5, Canadian National Defense, 2005).
- **U.K. Defence:** "Leadership is visionary; it is the projection of personality and character to inspire people to achieve the desired outcome. There is no prescription for leadership and no prescribed style of leader. Leadership is a combination of example, persuasion and compulsion dependent on the situation. It should aim to transform and be underpinned by individual skills and an enabling philosophy. The successful leader is an individual who understands himself/herself, the organization, the environment in which they operate and the people that they are privileged to lead" (p. 4, U.K. Defence Leadership Centre, 2004). The document adds that the key elements of the definition are "Vision", "Projection of Personality and Character", "Inspiring People", "Achieving the desired outcome", "Transforming", "An enabling philosophy", and "Self-awareness".
- **Singapore Armed Forces (SAF):** "Leadership is defined as a process of influencing people to accomplish the mission, inspiring their commitment and improving the organization" (p. 65, Chan et al., 2005).

BOX 1-2

"Leadership" in Academic Textbooks

- **Leadership: Enhancing the lessons of experience** (Hughes, Ginnett & Curphy, 2009): Emphasizing that "the central idea" of their textbook is that "leadership is a process, not a position", and recognizing that there is no single correct definition of leadership, Hughes et al. (2009) in the sixth edition of their textbook define leadership as "the process of influencing others towards achieving group goals" (p. 6).
- **Leadership in Organizations** (Yukl, 2010): Noting that "like all constructs in social science, the definition of leadership is arbitrary and subjective", Yukl (2010) provides the following as a "working definition" for leadership in his textbook which is into its seventh edition: "Leadership is defined as a process of influencing others to understand and agree about what needs to be done and how to do it, and the process of facilitating individual and collective efforts to accomplish shared objectives" (p. 26). Yukl adds that his definition of leadership is not limited to processes that necessarily result in 'successful' outcomes… (that his) focus is clearly on the process, not the person and they are not assumed to be equivalent" (p. 27).
- **Leadership: Theory & Practice** (Northouse, 2010): The following are central to the phenomenon of leadership: (a) it is a process or an interaction between leader and follower; (b) it involves influence ("the sine qua non of leadership"); (c) it occurs in a group or organizational context; (d) it involves goal attainment. Northouse offers this definition of leadership in the fourth edition of his textbook: "Leadership is a process whereby an individual influences a group of individuals to achieve a common goal" (p. 3).

FIGURE 1-1. **Leadership is about influencing people in situations to accomplish outcomes and purposes in a certain context.**

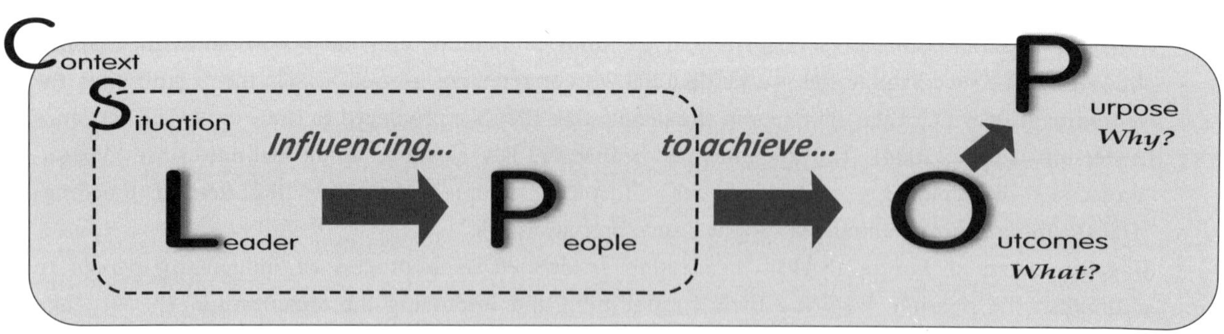

One can observe that there has been a convergence in the military and non-military thinking on leadership in recent times. This was partly due to the rapid growth in the social scientific knowledge of leadership in the last century during which a large number of research programs were initiated or supported with the help of military forces either in terms of research funding, samples, and operational demands or concerns. This convergence possibly came with the recognition that the concept of leadership has many meanings among laypersons who also seem to have biased understandings of the concept. It is common to find that even highly experienced military leaders often confuse the concept of leadership with related concepts such as management or command. For this reason, organizations like the military which deal with life or death situations care enough about the quality of leadership that they cannot leave this concept to individual interpretation. It is for this reason that modern military forces invest resources to clarify the concept of leadership as part of their doctrines.

Why Leadership Is Not Well Understood

Different Meanings and Understandings. In a well-established leadership textbook, Northouse (2007) remarks: "(Leadership) is much like the words democracy, love, and peace. Although each of us intuitively knows what he or she means by such words, the words can have different meanings for different people. As soon as we try to define leadership, we immediately discover that leadership has many different meanings" (p. 2). In another established textbook, Yukl (2010) suggests that the different approaches to defining leadership may reflect the different assumptions of the person defining the concept. For example, some may define leadership mainly in terms of the personal characteristics or "traits" of the leader because they assume that leaders are "born" rather than "made". Others may define leadership in terms of the behaviors and skills of the leader because they believe that such skills may be learned through experience. Yet others may define leadership in terms of the complex relationship between leader and follower because they think of leadership as a process of mutual influence. There are also those who assume that leadership is derived from the formal position or appointment of the person in an organization; that by being appointed a leader means that the person will lead others effectively. These different understandings of leadership is one reason why in the *Handbook of Leadership*, Stogdill (1974) famously remarked: "there are almost as many definitions of leadership as there are people who have attempted to define it" (p. 7). This is perhaps another reason why military forces — which employ large numbers of people usually from a wide range of demographic backgrounds — would invest resources in developing doctrines and definitions of leadership — to provide a common understanding so as to guide the practice of leadership.

Layperson "biases". Social scientists have discovered that laypersons have certain biases in the way they think about leadership which results in many of them asking very simplistic questions about leadership. Since the 1970s, psychologists have observed that people have a natural tendency to attribute more blame to the person's internal traits, motives and abilities, than to situational factors like the weather, bad luck, task difficulty, etc., when trying to explain what caused an outcome or behavior. This bias is called the Fundamental Attribution Error (Jones & Harris, 1967; Ross, 1977) and is particularly strong in more individualistic, Western cultures. In the 1980s, psychologists also observed a common tendency among people to "romanticize" the idea of leadership in society. For example, employers would often stress a preference for leadership than for followership when hiring new job candidates. Students are encouraged in schools to strive towards leadership positions, and educational systems often want to "produce" leaders. Similarly, there is a tendency to attribute the success or failure of football teams and orchestras to the "leadership" of the manager-coaches and conductors than to the players or musicians themselves. Psychologists also observed how their own leadership theories tend to place more focus on the leader rather than on the follower or the situation in the leadership process. In the 1980s, Meindl (1990; see also Meindl, Ehrlich, & Dukerich, 1985) first labeled this tendency to overemphasize the leader's responsibility for outcomes while reducing the responsibility of followers in *The Romance of Leadership*. Combining ideas from the Fundamental Attribution Error and the *Romance of Leadership*, Hackman and Wageman (2005; see also Hackman, 2002) proposed that there is a specific tendency to identify the leader as the main cause of collective performance which they called "leader attribution error". This bias is particularly pervasive because the leaders tend to be highly visible while the structural or contextual factors affecting group or organization tend to be less visible to observers. It is because of such biases that laypersons often ask simplistic but faulty questions about leadership like "do leaders make a difference?" and or "are leaders born or made?" Throughout this book, you will learn how research efforts to answer these simple questions often showed how the questions were themselves based on a biased understanding which tends to treat leadership and leaders as almost synonymous, when leadership is in fact not only a function of the leader.

Different approaches. Yukl (2010) notes that "questions about leadership have long been a subject of speculation, but scientific research on leadership did not begin until the 20th century" (p. 20). The focus of much research has been on the determinants of leadership effectiveness (e.g. traits, abilities, behaviors, sources of power, aspects of situation, etc.). Other researchers have tried to predict leader emergence, and "why some leaders are or become bad". Today, it is recognized that the scientific study of leadership has evolved through several different approaches or "research paradigms" as follows:

- **The trait approach**, which emphasizes the leader's attributes such as personality, values, motives, skills and competencies.
- **The behavioral approach**, which tries to determine "what leaders actually do" to be effective.
- **The situational or "contingency" approach**, which emphasizes how situational or contextual factors such as the characteristics of the followers, the nature of the work, mission or task (including difficulty, complexity, urgency, risk/danger), the organization (structure, climate and culture) or nature of the relationship (e.g., authority), affect the overall leadership process.

- **The "New Leadership" approach**, which includes transformational, charismatic and visionary leadership theories that focus on how the leader helps followers to align to organizational objectives through vision and values (see Bryman, 1992).
- **The integrative approach**, which refers to broad leadership theories that attempt to include many classes of leadership variables from leader and follower attributes to leader behaviors, the nature of "influence", and the situational/contextual variables.

Confused with related concepts. In his textbook, Yukl (2010) noted that "leadership" is a word taken from common vocabulary and is therefore used loosely and in many ways confused with somewhat similar words like management, authority, power, control, administration, and supervision. In an attempt to get at the "core" of the "leadership" concept, Antonakis, Cianciolo and Sternberg (2004) stated: "Most leadership scholars probably would agree, in principle, that leadership can be defined as the nature of the influencing process — and its resultant outcomes — that occurs between a leader and followers and how this influencing process is explained by the leader's dispositional characteristics and behaviors, follower perceptions and attributions of the leader, and the context in which the influencing process occurs. For us, a necessary condition for effective and authentic leadership is the creation of empowered followers in pursuit of a moral purpose, leading to moral outcomes that are guided by moral means" (p. 5). They observed that leadership needs to be differentiated from two related concepts — power and management — that are themselves often confused with leadership:

- **Power** refers to the leader's potential means to influence others. Sources of power include for example, position authority (e.g., military rank and appointment), or, referent power or the respect and admiration earned from one's followers.
- **Management**, which refers to actions like planning, organizing, staffing, and controlling the (human and non-human) resources in an organization or effort so as to achieve order, stability, efficiency and consistency in the accomplishment of goals.

What's one's standard or criterion of leadership? At times, discussions about leadership can get very challenging because each one of us may have very different standards or "criteria" to what constitutes "good" or "bad" leadership. Consider these questions: Is a military leader's effectiveness measured in terms of "winning or losing the battle"? What if a leader manages to influence his units to win the battle but at the expense of the soldiers' willingness to fight future battles with him? Should "leadership" be measured in terms of the military rank achieved by the leader, suggesting that seniority or positional power reflects leadership effectiveness? Or, should leadership effectiveness be measured in terms of the ability to rally the soldiers despite impossible odds, even if the battle is eventually lost?

The fact is that leadership criteria are multidimensional and multifaceted in nature. One's "criterion" of leadership can range from task accomplishment, to follower morale and commitment, to the extent that the leader influenced organizational change and improvement — all of which may be competing "goals" in the leadership process. This challenge of determining the criterion in any discussion on leadership may be called the "criterion problem" in leadership. When clarifying one's "criterion" of leadership, it is useful to remind oneself of the following:

- "Emerging as a leader" (leadership emergence) or "being appointed to leadership" is not the same as "being an effective leader". For example, think of the many politicians who "emerge" during elections by getting voted in only to be thrown out of office within a few months or years due to poor performance.
- The *perception* of leadership effectiveness is not necessarily the actual leadership *effectiveness* measured in terms of "tangible results". Some leaders may be good at creating an impression of effectiveness which is not in turn reflected in actual group or organizational results or outcomes.
- Leadership effectiveness may also depend on *who* is judging or perceiving the leader. Ask the leader himself or herself, his or her subordinates, his or her superiors/bosses, his or her clients or customers, his or her peers, and you will find very different views or impressions of the leader. This partly depends on the perceiver's understanding of "what characterizes effective leadership" or his/her "implicit theory of leadership".
- "Leadership effectiveness" is not always only defined by the task performance outcomes or "results". It may also be defined by quality or "value added", or in terms of how ethically the results were achieved.

"MILITARY" LEADERSHIP

Doctrinal (Military) versus Academic (General) Definitions

As seen in Boxes 1-1 and 1-2, most military forces today define leadership in a way that is similar to that found in general academic textbooks. In its doctrinal pamphlet on "leadership", the Australian Defence Force (2007) goes as far as to clarify: "The term 'military leadership' is somewhat misleading because it implies that there is a category of leadership that is peculiar to the military. The term has been incorrectly used by civilian commentators to describe the legitimate practice of command — a practice where subordinates (wrongly called followers) are compelled to obey lawful orders. Leadership as a concept is the same the world over, no matter how or where it is practiced. The principles of leadership do not change in the military any more than the principles of leadership change from one country to another. What does change is the context and or culture in which leadership is practiced." (p. 2–5). This idea — that leadership is essentially a general phenomenon and that it is the context that makes military leadership unique — is echoed in the Singapore military's "Leadership 24–7 framework". Figure 1–2 illustrates how the Singapore military depicts its concept of leadership in terms of general aspects of leadership (i.e., values, behaviors, styles and the "self") which form a "triangle", which are articulated in the context of Singapore military's unique mission and purpose, its operating environment, and the outcomes desired which is represented by a "circle" (see Chan et al., 2005 for an elaboration of the framework).

Context-free versus Context-specific Knowledge

In a recent review of the research conducted on military leadership, Wong, Bliese and McGurk (2003) noted that there are two ways in which military leadership is studied scientifically. They called these "context-free" and "context-specific" approaches to military leadership (note: these terms were first introduced by Blair & Hunt, 1986). One can therefore think of military leadership knowledge from these two perspectives:

- **"Context-free" knowledge**, which refers to knowledge gained from research studies that use military samples to test general leadership theories that are applicable to any type of organizations. Examples are the numerous studies of "transformational leadership" conducted with military samples — a concept and related theories that originated from ideas in the political or business world, but which have also gained popularity in military organizations (see e.g. Bass, 1998; Bradley & Charbonneau, 2004).
- **"Context-specific" knowledge**, is derived from theories or studies that focus on the unique characteristics of the military context, including its organizational structure, culture, technology, missions, types of personnel, etc. Developing such theories or conducting such research requires

FIGURE 1-2. **Singapore military's "Leadership 24-7" framework.**

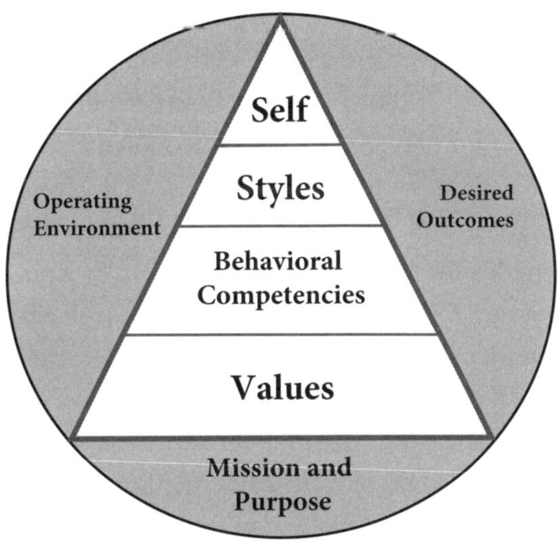

Source: POINTER: Journal of the SAF.

much more in-depth understanding of the military and the issues that it faces in different mission, organizational, personnel contexts. Wong et al. (2003) suggest that such studies are "likely to capture more of the true essence of what constitutes military leadership". They organized their "context-specific" review of military leadership research in terms of the nature of leadership at different levels of the military organizational hierarchy (e.g., direct, organizational strategic) and how the external environment (e.g., civilian society) and the mission context (e.g., garrison versus operational/field) that shape the critical tasks and demand certain individual/group capabilities and shape subcultures, thereby affecting the nature of military leadership.

Leadership versus Management

Although leadership and management are different concepts involving different foci and activities, both involve influencing groups of people to achieve goals. Yukl (2010) remarks: "A person can be a leader without being a manager (e.g., an informal leader), and a person can be a manager without leading. Indeed, some people hold the job title 'manager' but do not have any subordinates (e.g., a manager of financial accounts; p. 24)." The concepts of leadership and management also have different origins. While the concept of leadership was studied by many of the ancient philosophers such as Confucius (551–479BC) and Socrates (470–399BC) over 2,000 years ago, it was only about a century ago that the science of "management" first emerged with the publication of *Principles of Scientific Management* by Frederick Taylor (1911). Management was concerned with the efficient organization of work efforts and the reduction of chaos in organizations through principles such as scientific measurement of work, the standardization of tasks, and the division of labor. In 1916, the French engineer Henri Fayol published his 14 principles of management in which he also identified four key "management functions" including: planning, organizing, staffing, and controlling. Today, some military forces also define management with reference to these functions. For example, the U.K. Defence Leadership Centre (2004) defines management as "The allocation and control of resources (human, material and financial) to achieve objectives, often within the constraints of time. Management requires the capability to deploy a range of techniques and skills to enhance and facilitate the planning, organization and execution of business. (p. 5)."

In the 1970s and 1980s, some academic writers (e.g., Bennis & Nanus, 1985; Zaleznik, 1977) went as far as to argue that managers and leaders are different types of people (i.e., with different innate "traits"), suggesting that it was either not possible or desirable for people to play both roles at the same time. Bennis and Nanus (1985), for example, introduced their now oft-quoted phrase, "Managers are people who do things right and leaders are people who do the right thing" (p. 221), suggesting that managers and leaders have different goals or purposes. In this view, while managers strive to bring consistency, order and stability to organizations by implementing plans and procedures, leaders on the other hand strive to motivate people to act or to change on the basis of values and principles.

In a classic paper entitled *What leaders really do* and a book entitled *A Force for Change: How Leadership differs from Management*, Kotter (1990a, 1990b) argued that leaders and managers were distinct roles but these roles could co-exist in the same person in that leaders and managers deal with similar challenges but how they accomplish these is different. His key ideas are summarized in Table 1-1. Kotter's research however suggested that very few executives could carry out both roles effectively. He argued that the relative importance of leadership versus management depended on the situation — in a stable operating environment and as the organization becomes larger and more complex, the role of management increases. On the other hand, a more dynamic and uncertain external environment calls for more leadership than management. In his subsequent book entitled *What leaders really do*, Kotter (1999) stated that "Leadership and management are complementary systems of action. Each has its own function and characteristic activities. Both are necessary for success in an increasingly complex and volatile business world (p. 10)."

TABLE 1-1. Leaders versus Managers (adapted from Kotter, 1990a, 1990b)

Challenge	Managers (order and consistency)	Leaders (change and movement)
Complexity and change	Plan (set goals and targets) and budget (allocate resources).	Set a direction including a vision and strategies to achieve the vision.
Create capacity to achieve plans	Organize and staff the structure, communicate the plans, delegate responsibility and monitor performance.	Align people by communicating the new direction.
Ensure plan accomplishment	Monitor, control and solve problems.	Motivate and inspire; appeal to basic human needs, values and emotions.

Leadership versus Management in the Military. Just as academics started to differentiate the concepts of leadership and management in the 1970s and 80s, military writers also started to emphasize the difference in these concepts during this period. In a book that was highly critical of the state of U.S. Army leadership in the Vietnam War, Gabriel & Savage (1978) in *Crisis in Command: Mismanagement in the Army* argued that "as a consequence of adopting and implementing certain managerial and entrepreneurial practices while dismantling the more traditional modes of leadership behavior, levels of unit cohesion fell while overt indicators of unit disintegration rose at alarming rates" in the U.S. Army during the war. Today, the Australian Defence Force also stresses that "management and leadership are different concepts that generate different behaviors" (p. 1–1). It adds, "Leadership is generally seen as an influence process that energizes followers while management is a control process aimed at bringing coordination and efficiency to people and organizations". For the purpose of clarity and to avoid overlap within this doctrine, the activities of 'directing' and 'controlling' are seen as more to do with management than they are to do with leadership" (p. 1–6).

Leadership versus Command

Different Doctrines for Leadership and Command. Military forces introduce doctrines in order to provide some fundamental principles to guide soldiers' actions, in support of military objectives. While doctrines are authoritative, military forces also expect their personnel to excise judgment when applying such principles in different situations. It is therefore interesting to observe how established military forces today provide different doctrinal definitions of the concepts of leadership and command to guide their personnel. For example, in contrast to its definition of leadership as a "process of influencing people"; the U.S. military defines command as "The authority vested in an individual of the armed forces for the direction, coordination, and control of military forces" (see U.S. Department of Defense, 2001). The U.K. Defence Leadership Centre (2004) states that "Leadership… complements management; it does not replace it", and that "command requires the exercise of both leadership and management" (p. 8). The U.K. Defence Leadership Centre views leadership and management as "two sides of the coin of command", in that while management brings order and consistency to complex organizations involving planning and budgeting, leadership is concerned with setting direction, developing a vision of the future and strategies for achieving the vision. It illustrates this idea with an example: "Some commanders are poor leaders, but are able to function in the role by exercising managerial tools and positional power to get people to do the job. Their personal authority is not necessarily compromised during stable times. Yet, in a time of crisis, their shortcomings as leaders will be exposed by their inability to draw on the trust of their people and exercise influence. Commanders who can lead, but who lack the skills to manage, can still be successful if they compensate for their deficiencies by either recruiting staff who have the required skills or by personally developing" (p. 7). In this view, the U.K. Defence Leadership Centre stresses that "Leadership, to be effective, must extend beyond the authority of command. Thus, 'personal power' is the influence that inspires followers to follow beyond the transactional contract" (p. 8).

Is this leader exercising leadership, command or management?
Source: Robert Nickelsberq/Getty News Images/Getty Images

The "Human in Command". If the 1970s and 1980s witnessed a debate on the difference between the concepts of leadership and management in both the academic and military domains, it can be said that the 1990s witnessed a debate on the concepts of leadership versus command in the military world. In its Field Manual on "Mission Command" (FM 6-0), the U.S. Army (2003) described how the idea of "command and control" as opposed to simply command first entered Army terminology during World War II, and as warfare became more complex, "command and control" became considered a "battlefield operating system" in the late 1970s related to conduct of operations involving planning, preparation, execution, and assessment. With the end of the Cold War and an apparent rise in low-intensity type of conflicts and peace support missions in the 1990s, military forces became more concerned about the human challenges faced by commanders in the new military operating environment, that it was no longer adequate to think of "command" only as part of "command and control systems". In June 1998, a NATO workshop was organized entitled the *Human in Command* which marked a shift from what was an increasingly technical-systems focus on "command and control", to a more human-centric focus on the experience and "art" of command (McCann & Pigeau, eds., 2000). One of the more profound realizations at this conference was the idea emphasized by Canadian military behavioral scientists that "only humans command", that "control" is a tool of command (i.e., those technical systems, structures and processes that are devised by "command" to enable "command" and to manage risk), and that it is commanders that must create, initiate and change "control", as opposed to control driving how commanders command (see Pigeau & McCann, 2002). In 2003, the U.S. Army published a new doctrine in its Field Manual 6–0 entitled *Mission Command: Command and Control of Army Forces* that stressed that the "principal aspect" of command and control is "command", that command is an art while control is the "science" that supports the commander to command amidst the uncertainties of the battlespace. The following were identified as the "elements" of command: Authority, decision-making and leadership. The new U.S. Army doctrine stressed that commanders should combine the art of command and the science of control to exercise "command and control"; and that to do this, military commanders should employ the behaviors such as "visualizing the battlespace", "describing their vision to subordinates", "directing actions to achieve results", and "leading the command to mission accomplishment". Therefore, this approach — just like the U.K. military — would treat leadership as a part of the overall process of command. Other militaries such as the Australian and Canadian forces however emphasize the difference among the three concepts in their leadership doctrines — perhaps to highlight to their personnel the importance of leadership *beyond* command and management.

Australian and Canadian Forces on Command, Leadership and Management. Today, the Australian Defence Force (2007) states, "It is obvious that the essence of command is the legal authority to order subordinates toward assigned tasks. Command requires structure, can only be applied down the chain of command and requires compliance from followers. NATO's definition places primacy on accomplishing the mission with a secondary focus on the personnel involved. The definition doesn't include the word 'leading', nor does it mention that vital component of leadership, namely the influence of others" (p. 2–3). The Canadian Forces (2005) also clarifies the distinction between leadership, command and management in its doctrine as follows: "Leadership is an essential role requirement for commanders and managers but is not the same thing as command or management. Command is bounded by the lawful authority delegated to a commander and may only be exercised down the chain of command. Management is similarly bounded by the authority delegated to a manager and may only be exercised down the managerial chain of command. Leadership, on the other hand, may be exercised by anyone, regardless of organizational position" (p. 7).

From the above discussions, readers should now better understand the meaning of leadership as it is used in various military forces — relative to the concepts of command and management (which are often confused with leadership). Figure 1-3 below summarizes some of the main ideas in the concepts of leadership, command and management, while emphasizing how these three concepts are related and yet different from one another. To the extent that these are three concepts that have no absolute meaning — it is up to each military to spell out how it wishes its personnel to understand and behave in relation to each concept.

FIGURE 1-3. Leadership, command and management in the military context.

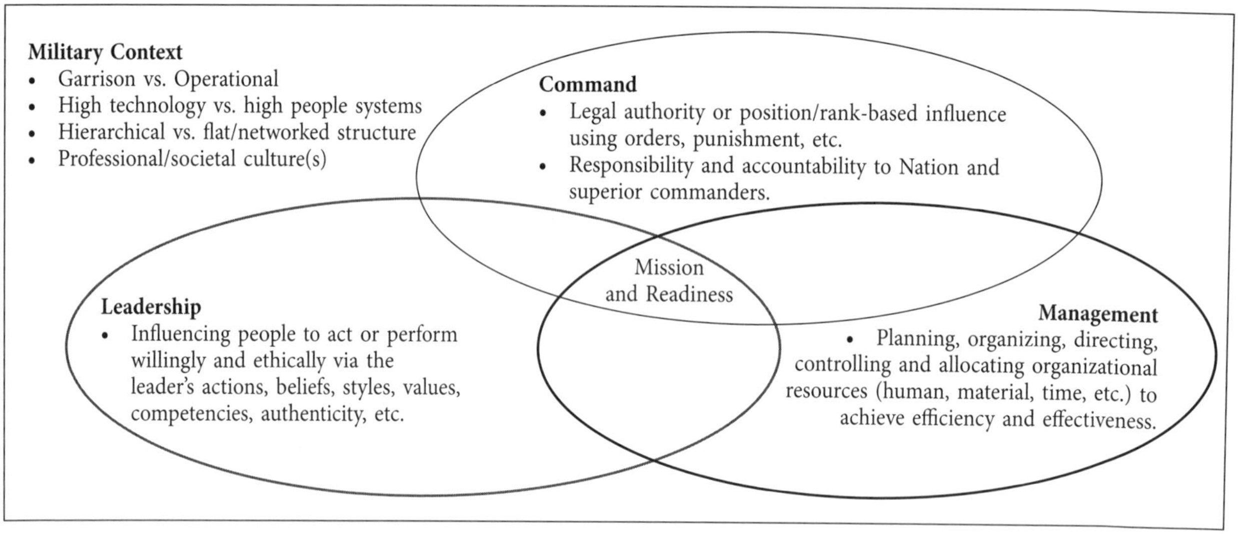

THE PRACTICE OF MILITARY LEADERSHIP TODAY

Why Military Leadership Matters

Writing for a non-military audience, Kotter (1990) once argued: "Consider a simple military analogy: A peacetime army can usually survive with good administration and management up and down the hierarchy, coupled with good leadership concentrated at the very top. A wartime army, however, needs competent leadership at all levels. No one has yet figured out how to manage people effectively into battle; they must be led" (p. 40). This is at best an intuitive argument for why leadership matters in the military. How do military forces argue for the importance of leadership in their business or profession? The following themes seem to emerge when one engages military leaders on the subject.

The Human Factor: Military Morale and Motivation. U.S. Army war-fighting doctrine (Field Manual 3-0, 2001) states that Combat Power or a military force's ability to fight is created by combining various elements such as maneuver, firepower, protection and leadership, where leadership is seen as the "most essential dynamic" because it is leadership that inspires the soldiers by providing the purpose, direction and motivation that is required to win and to persist in battle. Emphasizing its importance in maintaining combat motivation, the New Zealand Defense Forces (2008) states: "Leadership at all levels is the principal element in the maintenance of morale. Without good leadership, morale will undoubtedly crumble in the face of adversity. All leaders must accept their responsibility for maintaining morale and the fighting spirit of those under their command" (p. 7–7).

Military leaders are responsible for the morale and motivation of soldiers under their command.
Source: CyberPioneer, Ministry of Defence, Singapore.

For Operational Readiness and Relevance. Intrinsic to the military profession is a paradox that soldiers are constantly training to prevent the very reality that they train for — war and conflict. Besides being key to operational success, military leadership is also the key to maintaining military forces at a high state of operational readiness. This is done by constantly developing and improving military doctrine, and by training the troops to prepare for new and emerging threats. Military leadership must balance the need to prepare the soldiers to deal with a wide range of known and unknown operational challenges while maintaining a high state of morale, commitment and cohesion among the troops.

Maintenance of Military Ethos and Civil-military Relations. When compared with other professions like medicine, law or teaching, the military is unique in that its only "client" is the country that it protects. The military is also conferred the power to use force to kill and destroy for specific reasons sanctioned by the State, and military personnel are expected to risk their lives to achieve these missions if necessary. Because of the "high stakes" nature of the military business, military leadership plays a vital role in maintaining the strong professional ethics that enables the troops to perform their missions, while also maintaining public trust in the way the military handles its business.

What Shapes Doctrines (that guide the practice of military leadership today)?

Operational Experiences and "Lessons learned". It is possible to see how the U.S. military derived its basic principles related to military leadership from its experience of wars. For example, in a large scale research study reported in four volumes called *The American Soldier*, Stouffer et al. (1949) surveyed the attitudes, beliefs and experiences of over half a million of U.S. soldiers during World War II. They developed and employed various socio-psychological techniques including surveys, interviews and observations. For example, they asked soldiers "what kept them going during the war" and found that the most common responses were "getting the war over with so that they could go home" and "you couldn't let the other men down". From various studies using different methodologies, Stouffer et al. concluded that ideology, patriotism, or fighting for the cause were secondary to cohesion or the emotional ties between soldiers which were the primary factor in combat motivation. Other influential World War II studies that have shaped modern military leadership thinking include Shils and Janowitz's (1949) study that reported on the importance of small unit leadership and cohesion in affecting the morale and motivation of the German soldiers; and, S.L.A. Marshall's 1947 classic, *Men against Fire*. More recent "experiences" that have shaped the scientific knowledge of military leadership include the analysis of the U.S. military's leadership challenges in the Vietnam War (e.g., Gabriel & Savage, 1977; Henderson, 1985), and the study of factors that led to the disastrous shooting down of a civilian airliner by the USS Vincennes in 1988 (see U.S. Navy research program called Tactical Decision Making Under Stress or "TADMUS"; Cannon-Bowers & Salas, 1998), and a recent set of studies of "military life" edited by Britt et al. (2007).

New/Future Missions, Systems and Operating Environments. Modern military forces operate on the principle of not preparing for yesterday's wars. Instead, they constantly anticipate future military missions and systems and how these may impact the nature of leadership. Many of such efforts to anticipate the future of military leadership involve highly specialized behavioral science units and personnel in the various forces. Regular and ad-hoc networks and meetings are also organized by the various military forces and research agencies to discuss and share knowledge on issues related to the future challenges and issues related to military leadership (see Zaccarro, Klimoski & Boyce, 1999). Some international conferences that regularly discuss contemporary military leadership issues are the International Military Testing Association (IMTA) conference, the International Applied Military Psychology Symposium (IAMPS), and the Inter-University Seminar (IUS) on Armed Forces and Society. Some notable meetings that shaped the current North American and NATO military leadership doctrines include a 1996 Army Leadership Symposium held at Cantigny Estate, Wheaton, Illinois entitled "Leadership Challenges of the 21st Century Army" (see Hunt & Phillips, 1996). The discussions at the Symposium resulted in a book entitled *Out-of-the-box-leadership: Transforming the Twenty-first-century Army and other top-performing organizations* (Hunt, Dodge & Wong, Eds., 1999) which anticipated many of the changing military missions in the 21st century. Another important meeting was organized by NATO's Research and Technology Organization that subsequently resulted in a book called *Human in Command: Exploring the Modern Military Experience* (McCann & Pigeau, Eds., 2000), which till today, influences military doctrines on command versus leadership.

Personnel and Demographic Changes. Military leadership doctrines are also shaped by the structural and personnel changes in the organization which are often a function of changing societal demographics and recruitment/retention factors. Recognizing changes in the ethnic composition, education and expectations of its personnel, the Canadian Forces (2005) wrote in its *Conceptual Foundations of Leadership* manual: "In addition to adapting to new operational requirements, CF leaders must also adjust to the changing characteristics of the recruit pool and the equally shifting composition of the armed forces. . . . to be effective in building individual commitment and teamwork, leaders will have to demonstrate strong interpersonal skills, a willingness to be open and candid with their subordinates, and behavior that is both mindful of differences in maturity and conscientious in addressing individual and group concerns" (p. xiv).

THE STUDY OF MILITARY LEADERSHIP

Different Approaches

Historical. Bass (1990) remarked that "The study of leadership rivals in age the emergence of civilization, which shaped its leaders as much as it was shaped by them. From its infancy, the study of history has been the study of leaders — what they did and why they did it (p. 3). Military history, or the study of past military campaigns and battles, is an important way to understand the importance of military leadership. This is done either by describing or analyzing the actions and decisions of a particular commander or command group, or by comparing the decisions and actions of military leaders and commanders on the different sides, and how these influenced the outcomes of the operations. With its focus on the events that occurred in time, the historical approach also allows one to learn about "leadership strategies", or "how to think and act under certain circumstances". The historical approach helps us to appreciate that military leadership is important not only in terms of the moment by moment actions and decisions made by the leader in battle, that it is also important in the entire preparation and organization of the military forces and in the conduct of operations. The strength of the military history approach is that it enables us to appreciate the context in which the leaders acted as they did. "Context" includes many factors such as the quality and training of the troops, how the troops were organized or deployed, the available military technology and how it was employed, the weather and unfolding of events, and the commanders' application or adherence to the principles of war. There may be little that a leader can do to salvage poorly trained or organized soldiers. A limitation of the historical approach is that just as most people have certain biases in their understanding of leadership (e.g., leadership attribution error), it is possible that some historians may in their analyses attribute more blame to leaders than to the contextual circumstances that led to operational failure.

Military leadership is traditionally studied from a historical perspective.
Source: Robert Alexander Hillingford/The Bridgeman Art Library/ Getty Images.

Experience, Wisdom and "Authority". Ultimately, leadership is not learned from any textbook, course or "list of principles". Rather, one learns to lead through actual experience. Such experiential learning is enhanced if the experience is complemented with reflection on the knowledge of leadership from history, the wisdom of "great" or respected leaders, or social scientific theories of leadership. One can therefore find many popular business books attempting to proffer "leadership advice" to managers. An example is the former U.S. Chief of Staff, Joint Staff General Colin Powell's "Rules of Leadership" which is easily found on the Internet and often cited by many leadership speakers (e.g., Harari, 2002). Such principles or lists make more sense when one makes the effort to understand the historical context in which they "worked", and they serve to guide one when dealing with similar circumstances in the present. Many of these writings make it to the bestseller list because the principles are accompanied by stories told by the leaders who were actually "there". One can learn from these leaders especially if one is familiar with them or respects them tremendously. However, whether these principles are internalized depend on one connecting the leader's wisdom with one's own experience of leading.

Social-scientific. Leadership is studied across various disciplines in the humanities and social sciences including psychology, sociology, political science and education. As with the physical sciences, the social sciences are approaches to knowing that are based on observation and evidence. This is in contrast to knowledge from authority (e.g., religion), tradition or custom, pure logic or reason (e.g., philosophy). Scientists usually approach any topic of study with a dose of skepticism — they constantly think to themselves: "I observe this, people seem to think this, but is this true?" Scientists will develop theories and hypotheses logically and by observing patterns and associations among variables to determine cause and effect, and subject these to empirical tests based on observable, measureable evidence where possible. Scientific knowledge is also cumulative and integrative; scientists will revise their theories with new evidence and considerations. Today, it is increasingly common to find "meta-analytic reviews" that aggregate or average the statistical findings across numerous studies while correcting for statistical artifacts and measurement errors to provide more stable estimates of relationships, including cause and effect or correlational ones. Scientific knowledge is also shared publicly through academic publications. Scientific research can be more "basic" or "applied" in nature. Basic research refers to studies that aim to test fundamental theories for the sake of evaluating them. In contrast, applied research is driven by practical questions or problems that need a solution. Much of the social scientific knowledge of leadership described in this book derives from both basic research on leadership-related theories and concepts (e.g., to test a theory of leadership "styles"), and from real-world military problems and needs, for example, to develop better assessment tools for use in leadership selection and/or development. Some military research has led to the development of new social science theories or concepts. For example, the *American Soldier* studies by Stouffer et al. (1949) were not only important for their findings on military psychology and leadership, they also contributed to basic knowledge related to the psychology of leadership, for example, in the sociological concept of "relative deprivation", which refers to the discontent or dissatisfaction that people feel when they compare their positions or statuses to those around them and find that they have less than their peers. In many ways, military needs have driven advances in scientific research and development. Applied behavioral scientists also seek to improve individual, group and organizational work processes through the application of systematic procedures and research-based principles. Many psychological tools and techniques used in leadership selection today were also first developed for military use. For example, the first measures of general intelligence designed for mass testing, the Army alpha and beta, were developed to meet the U.S. Army's screening and personnel selection needs during World War One from 1917 to 1918. It was the U.S. Office of Naval Research (ONR) that funded Ralph Stogdill's "Ohio State Leadership Studies" that shifted the focus of leadership research from traits to behaviors. The U.S. ONR also funded the development of the "Critical Incident Technique" by Flanagan (1954) to evaluate naval aviators. This technique is now widely used in human resources research to analyze job requirements. The technique of "peer ratings" was also first developed by Edwin Hollander (1954) working as a Naval Aviation Psychologist at Pensacola in the early 1950s, where he showed that peer nominations could predict successful completion of Naval Air Training. More recently in 2010, the U.S. National Defense University published a book entitled "*The 71F Advantage: Applying Army Research Psychology for Health and Performance Gains*" (Bartone et al, 2010), highlighting the unique contributions of U.S. Army Research Psychologists with an entire section dedicated to the psychology of leadership.

Why Study the Science of Military Leadership?

A Living and Advancing Science. One key characteristic of science is the cumulative nature of its knowledge, where "new" findings are also evaluated in terms of validity and relevance (and boundaries) against existing knowledge in the field on the basis of empirical evidence. In a statement to the U.S. House Committee on Armed Services, the leading military sociologist, David Segal (2008), summarized the contribution of the social science and behavioral sciences to the military in the last century in the areas of maximizing soldier and unit performance, leadership and vocational selection

21st century operations present many new challenges for military leadership.
Source: CybePioneer, Ministry of Defence, Singapore.

and classification, understanding the factors that affect soldier morale and performance, and in helping military forces to transit from conscript-based to all volunteer force structures. In the same year, a report to the U.S. National Research Council on *Human Behavior in Military Contexts* (Blascovich & Hartel, Eds., 2008) summarized the state of scientific knowledge about leadership as follows: "Research over the past 100 years has shown that the key elements of credibility are task-relevant competence and trustworthiness. Leaders need to develop relationships with followers, both individually and as team members, which motivate and enable them to contribute maximally to mission accomplishment." Clearly, one important reason for studying the science of military leadership is the vast amount of careful social scientific research conducted on the subject in the past century. While the scientific knowledge of military leadership has witnessed an explosive rate of growth in the last century, one can expect that this field will continue to grow further with increasing recognition of the work carried out by increasing numbers of military social scientists and with increasing non-military interest in understanding the nature of leadership in the dynamic operating environments fraught with risk and uncertainty. This trend is evident in a recent special issue of the journal *Military Psychology* that has brought together theories and research on the topic of "Leadership in Dangerous Contexts" (Campbell et al., 2010) and a call for more research on subject (see also Hannah et al., 2010). Non-military interest in the study leadership in the modern military context can also be seen in the *Harvard Business Review Magazine* which featured several articles in its November 2010 issue that highlighted to the business world the challenges of leading in modern military operations.

Scientific Principles Endure. This book attempts to demonstrate to readers the scientific basis for many ideas that are found in contemporary military leadership doctrines. One reason why modern military forces base their doctrine on the social sciences is because science attempts to establish principles that can generalize across populations and times, and in doing so, also spells out the boundaries of what is known and not known about different phenomena. Writing for a professional military audience, Bass (1998) stated in the U.S. Army's professional journal, *Military Review*: "The principles of leadership do not change; only the conditions in which they are applied. Over time, we gain a better and more accurate understanding of the concepts and principles, but they were in effect like Initiation and Consideration when Julius Caesar exhorted his troops before storming Alesia — although it took us 1,900 years to refine and measure the concepts" (p. xx). A scientific approach to "military leadership" also helps to ensure great consistency of understanding and standardization of actions related to the many "fuzzy" ideas related to leadership.

For the Practice of Adaptive Military Leadership. Educating military personnel on the social scientific principles of leadership rather than merely providing procedures or tools can generate future military leaders who can "think on their feet" amidst the increasingly uncertain and dynamic nature of military missions. In 2005, the Canadian Forces not only published a "Leadership Doctrine" manual for its general military audience, but also published a manual on the "Conceptual Foundations" for its leadership doctrine. It explained this emphasis on frameworks, theories and principles as follows: "While drills and standard procedures work superbly in routine task environments and predictable situations, they tend to limit effectiveness in novel problem situations and in disorderly unpredictable environments. In these latter circumstances, an understanding of principles and concepts is the key to creative and adaptive behavior" (p. ix).

Providing a "Systems View" of Military Leadership. The social-behavioral science approach has moved away from "Great Man" assumptions of leadership to look at leadership as a process of influence between leaders and followers in the context of the mission, task, and operating and/or organizational condition. Hence, in contrast to studying the actions and decisions of a single military general or "Great Leader", the social-behavioral science of military leadership encourages a more systems view of military leadership in terms of behaviors that occur "in context". From an applied perspective, the Canadian Forces (2005b) also adds that "Because of the emphasis on the antecedents, behavioral dimensions, and consequences of effective CF leadership, it should also be of particular interest to staff involved in leader selection, development, and evaluation" (p. ix). Writing in *Joint Forces Quarterly*, Price (1996) noted: "Having the right leader in the right place at the right time has always been critical to victory. Yet for an institution such as the Armed Forces it is not leaders who bring success but leadership. Is there a real difference between individual leaders and institutional leadership? History is replete with examples of the importance of a leader's personal character, courage, and skill to the outcome of an operation. But unlike battles, campaigns, or even protracted conflicts, military institutions are long-term organizations which have enduring political, cultural and social values. Their key to success is not a single outstanding leader or even a succession of them. Institutional achievement is founded on a system of ongoing collective leadership that transcends individuals" (p. 97). In this sense, it is important for all personnel (including the seniors or "Generals") in modern military forces to adopt a more systems view of leadership as a process, rather than to view leadership in terms of the leaders as persons or leadership as a position."

SUMMARY

Contemporary military definitions of leadership are very similar to the definitions adopted in general texts — they describe leadership as a process of influence. In this approach, the essence of leadership is the process of influence involving leaders and followers "in context". The practice of leadership is not only about the leaders or persons who hold leadership positions. In the general academic context, leadership is distinguished from related concepts like management and power while in the military, leadership usually is treated as doctrinally distinct from command — it may be one aspect of "command", but is not synonymous with "command". If command is associated with positional power and authority that is based on law, leadership is associated with influence based on personal or referent power — beyond the authority that comes with rank and command. This very distinction becomes important in appreciating the unique meaning and practice of leadership in modern military forces.

There are different approaches to the study of military leadership. These include the historical, biographical and social-scientific approach, each with its own strengths and limitations. This book adopts the social-behavioral science approach because this often provides the basis for many contemporary military doctrines, including those of the U.S., Canadian, Australian, U.K., New Zealand and Singapore military. Understanding leadership is a challenge for various reasons — most people have their own notions of what leadership means and there are also natural biases in how people think about leaders and leadership. Leadership is also a complex phenomenon (e.g., criterion problem). Military forces place great emphasis on leadership because it is critical to operational success — both in the present and for the future. Hence, modern military forces today invest heavily in defining, clarifying and educating their leaders on their "doctrines" and fundamental principles of leadership, both to guide its practice and for systematic leadership development at the individual and organizational levels, and for more adaptive — as opposed to rule-bound — leadership.

Key Concepts

- Leadership
- Management
- Command
- Power
- Criteria of leadership
- Military Doctrine
- Theory and concepts (scientific)
- Romance of Leadership
- Fundamental/Leadership Attribution Error
- Context-free approach/knowledge
- Context-specific approach/knowledge
- Scientific approach
- Social sciences
- Approaches to leadership

Questions

- How does the military in your country define or discuss the concept of leadership? What is its basis for its leadership doctrine (i.e., is your military's thinking on leadership based on science or history or a certain philosophy?)
- Think of a military leader you know and evaluate this leader in terms of his/her ability to lead, command and manage. Do you find it useful to distinguish the concept of leadership from similar concepts like command and management?
- Why are you studying military leadership? What do you hope to do with this knowledge — both for your own leadership and for other leaders that you may develop or lead?

References

Antonakis, J., Cianciolo, A.T. & Sternberg, R.J. (Eds.). (2004). *The nature of leadership*. Thousand Oaks: Sage Publications.

Australian Defence Headquarters (2007). *Leadership in the Australian Defence Force* (ADDP 00.6). Director Centre for Defence Leadership Studies, Defense Publishing Service, Canberra, ACT.

Bartone, P.T., Pastel, R.H., Vaitkus, M.A. (Eds., 2010). *The 71F advantage: Applying army research psychology for health and performance gains*. Washington, DC: National Defense University Press.

Bass, B.M. (1990). *Bass & Stogdill's handbook of leadership* (3rd ed.). New York: The Free Press.

Bass, B.M. (1998). Leading in the Army After Next. *Military Review, 78(March),* 46–57.

Bennis, W. & Nanus, B. (1985). *Leaders: The Strategies for Taking Charge*. New York: Harper & Row.

Blascovich, J.J. & Hartel, C.R. (eds., 2008). *Human behavior in military contexts*. Washington, DC: National Research Council.

Bradley, J.P. & Charbonneau, D. (2004). Transformational leadership: something new, something old. *Canadian Military Journal, 5(1)*, 7–14.

Britt, T.W., Adler, A.B., & Castro, C.A. (eds., 2005). *Military life: The psychology of serving in peace and combat [Four Volumes]*. Praeger Publishers

Bryman, A. (1992). *Charisma and leadership in organizations*. London, Sage.

Campbell, D.J., Hannah, S.T., & Matthews, M.D. (2010). Leadership in military and other dangerous contexts: Introduction to the special topic issue. *Military Psychology, 22,* 157–189.

Cannon-Bowers, J.A., & Salas, E. (eds., 1998). *Making decisions under stress: Implications for individual and team training*. Washington, DC, US: American Psychological Association.

Canadian National Defense (2005). *Leadership in the Canadian Forces*. Canadian Forces Leadership Institute, Canadian Defense Academy.

Chan, K., Singh, S., Ramaya, R., & Lim, K. H. (2005). *Spirit & System: Leadership Development for a Third Generation SAF [4th Pointer Monograph]*. Singapore: SAFTI Military Institute.

Fayol, H. (1930), *Industrial and General Administration* (translated by Coubrough, J.A.). London: Sir Isaac Pitman & Sons.

Flanagan, J.C. (1954). The critical incident technique. *Psychological bulletin, 51(4)*, 327–358.

Gabriel, R.A. & Savage, P. (1978). *Crisis in Command: Mismanagement in the Army*. New York: Hill & Wang.

Hackman, J.R. (2002). *Leading teams: Setting the stage for great performances*. Boston: Harvard Business School Press.

Hackman, J.R., & Wageman, R. (2005). When and how team leaders matter. *Research in Organizational Behavior, 26,* 37–74.

Hannah, S.T., Campbell, D.J, & Matthews, M. D. (2010). Advancing a research agenda for leadership in dangerous contexts. *Military Psychology, 22,* 1–14

Harari, O. (2002). *The leadership secrets of Colin Powell*. NY: McGraw Hill.

Henderson, W.D. (1985). *Cohesion: The human element in combat*. Washington, D.C.: National Defence University Press.

Hollander, E.P. (1954) *Buddy ratings: Military research and industrial implications*. Personnel Psychology, 7, 385–393.

Horn, B. and Maclntyre, A.T., (eds., 2006). In Pursuit of Execellence: *An Overview of International Military Leadership Centres*. Kingston: CDA Press.

Hughes, R.L., Ginnett, R.C., & Curphy, G.R. (2009). *Leadership: Enhancing the lessons of experience* (6th ed.). Boston, MA: McGraw Hill.

Hunt, J.G. & Blair, J.D. (eds., 1985). *Leadership on the future battlefield*. New York: Pergamon Press.

Hunt, J.G., Dodge, G.E., & Wong, L. (eds., 1999). *Out-of-the-box-leadership: Transforming the Twenty-first-century Army and other top-performing organizations*. Stamford, CT: JAI Press.

Hunt, J.E., & Phillips, R.L. (1996). *1996 Army symposium: Leadership challenges of the 21st century army: Executive summary*. (Research Note 96–63). Alexandria, VA: United States Army Research Institute for the Behavioural and Social Sciences.

Jones, E.E. & Harris, V.A. (1967). The attribution of attitudes. *Journal of Experimental Social Psychology, 3,* 1–24.

Kotter, J.P. (1990). What leaders really do. *Harvard Business Review, 68*, 103–111.

Kotter, J.P. (1990). *A force for change: How leadership differs from management*. New York: Free Press.

Marshall S.L.A. (1947). *Men against fire: The problem of battle command in future war*. New York: William Morrow.

McCann, C., & Pigeau, R. (2000, Eds.). *The Human in command: Exploring the modern military experience*. NY: Kluwer Academic.

Meijer, M., & de Vries, R. (2005). Maintaining combat readiness in the Royal Netherlands Armed Forces: The psychosocial perspective. In *Strategies to Maintain Combat Readiness during Extended Deployments — A Human Systems Approach* (pp. 10–1 – 10–8). Meeting Proceedings RTO-MP-HFM-124, Paper 10. Neuilly-sur-Seine, France: RTO.

Meindl, J.R. (1990). On leadership: An alternative to the conventional wisdom. Research in Organizational Behaviour, 12, 159–203. 45 pages.

Meindl, J.R., Ehrlich, S.B., & Dukerich, J.M. (1985). The romance of leadership. *Administrative Science Quarterly, 30,* 1, 78–102.

New Zealand Defense Force (2008). *Foundations of New Zealand military doctrine* (NZDDP–D, Second Edition). Wellington, New Zealand: Headquarters New Zealand Defence Force.

Northouse, P.G. (2007). *Leadership: Theory & practice* (4th ed.). Thousand Oaks, CA: Sage.

Price, David E. (1996). Leadership: Some thoughts on the military circa 2025. *Joint Forces Quarterly, 13(3),* 97–99. Available at: http://www.dtic.mil/doctrine/jel/jfq_pubs/2013.pdf

Pigeau R., &, McCann C. (2002). Re-conceptualizing command and control. Canadian Military Journal, 3(1), 53–63.

Ross, L. (1977). The intuitive psychologist and his shortcomings: Distortions in the attribution process. 'In L. Berkowitz (Ed.), *Advances in experimental social psychology* (vol. 10, pp. 173–220). New York: Academic Press.

Segal, D.R. (2008). *Statement of David R. Segal to Subcommittee on Terrorism, Unconventional Threats and Capabilities*. Washington, DC: House Committee on Armed Services and Subcommittee on Research and Science Education. Retrieved from http://gop.science.house.gov/Media/hearings/research08/april24/segal.pdf

Shils, E.A., & Janowitz, M. (1948). Cohesion and disintegration in the Wehrmacht in World War II. *Public Opinion Quarterly, 12,* 279–315.

Stogdill, R.M. (1974). *Handbook of leadership: A survey of theory and research*. NY: Free Press.

Stouffer, S.A., Suchman, E.A., DeVinney, L.C., Star, S.A., & Williams, R.M. (1949). *Studies in Social Psychology in World War II: The American Soldier Vols. 1–4.* Princeton: Princeton University Press.

Taylor, F.W. (1911). *Principles of scientific management.* New York & London: Harper & Brothers.

U.K. Defence Leadership Centre (2004). *Leadership in defense.* Defense Academy of the United Kingdom, Shrivenham, Swindon.

U.S. Army (2001). *Operations.* (FM 3–0). Washington, D.C.: Headquarters, Department of the Army.

U.S. Army (2003). *Mission Command: Command and Control of Army Forces.* (FM 6-0). Washington, D.C.: Headquarters, Department of the Army.

U.S. Army (2006). *Army leadership: Competent, confident, and agile.* (FM 6–22). Washington, D.C.: Headquarters, Department of the Army.

U.S. Department of Defense (2001). *Dictionary of military and associated terms* (Joint Publication 1–02). Washington, DC.: The Pentagon.

Wong, L., Bliese, P., & McGurk, D. (2003). Military leadership: A context-specific review. *The Leadership Quarterly, 14(4),* 657–692.

Yukl, G. (2010). *Leadership in organizations (7th ed.).* Boston, MA: Prentice Hall.

Zaccaro, S.J., Klimoski, R.J., & Boyce, L.A. (1999). *The changing U.S. Army: A summary of future focused reports from 1990–1999* (RR-99-1747). Alexandria, VA: U.S. Army Research Institute.

Zaleznik, A. (1977, May–June). Managers and leaders: Are they different? *Harvard Business Review, 55,* 67–78.

2

Challenges and Stress in Military Operations

Source: MCT/McClatchy-Tribune/Getty Images

OVERVIEW

The previous chapter introduced the social-behavioral sciences approach to the study of military leadership in which the knowledge formed the basis for leadership doctrines and practices in many modern armed forces today. Before discussing questions such as "what are the attributes of an outstanding leader" or "what behaviors and styles should a leader adopt", it is important that military leaders first understand the context and its demands that impact them and their leadership. The context refers to the environment, tasks and the soldiers that the leader has to work with. Knowledge of the context will facilitate a deeper appreciation of the studies and findings on leadership. Therefore, this chapter and the next two focus on the context of leadership.

This chapter highlights the key challenges and stressors of modern operations and the battlefield, the resulting impact on soldiers and the leader, and what military forces do to mitigate the impact. The first section highlights the changes that took place in the last few decades and examines the resultant five key challenges that most commanders have to face today, if not, in the future. Each challenge and its implication for the commander are discussed. The subsequent sections focus on the stressors and combat stress during operations (e.g., Can fear be controlled? How does stress affect the body and performance? How does killing affect the soldier's mental health?) and how to manage the stress (e.g., What should a commander do to minimize combat stress? How can a commander build mental resilience in his/her troops?).

CHALLENGES FACING COMMANDERS TODAY

Thirty years ago, the main role of an infantry platoon commander would be to operate closely with his/her company and battalion in support of a mission to capture or defend a piece of territory against another military force. He/she would rely on the compass for navigation, flares to illuminate the killing area and request for company or battalion fire support on enemy soldiers. Today, the same platoon commander may be operating independently to enforce security or assist in humanitarian aid within an urban or suburban place of operation, but also needs to interact frequently with civilians, fight an enemy who does not wear a uniform, constantly remain vigilant against improvised explosive devices, and face the possibility of his/her actions being recorded or screened live for the world to witness. He/she may be equipped with a Ground Positioning System for navigation, night vision device to operate throughout the night, caffeine pills to stay alert and networked communications beyond the battalion to call for fire or download intelligence information. Without over generalizing, the above description of the military context illustrates the changes and challenges that modern military forces have experienced over the years. The rest of the section elaborates on the challenges and the implications for leaders today. There are a number of interesting books by several authors (e.g., Hedges & Al-Arian, 2008; Jordan, Kiras, Lonsdale, Speller, Tuck & Walton, 2008; Thussu & Freedman, 2003; Van Creveld, 2008) who discuss in greater depth the ideas discussed in this section.

The Challenge of Multiple and Complex Security Threats

In the past, warfare was characterized by armed confrontation between countries involving small to large scale military forces. With the collapse of the Soviet Union in 1991, irregular warfare has dominated the post cold-war era. Irregular warfare (IW) is "marked by a violent struggle among state and non-state actors for legitimacy and influence over the relevant population. IW favors indirect and asymmetric approaches . . ." (U.S. Department of Defense, 2006, p. xi). IW is most evident in the insurgents' struggle against the U.S.-led coalition occupation in Iraq and Afghanistan. In the last two decades, threats from failed states and ethnic conflicts (e.g., former Yugoslavia, Bosnia, Angola, Somalia and Sudan), transnational terrorism (e.g., the 9/11 attacks on the World Trade Center in New York City, the 2008 Mumbai attacks, and hijacking of ships in the waters off Somalia) and natural disasters (e.g., the 2004 tsunami, disaster in Indonesia, the 2010 earthquake in Haiti and more recently, the tsunami and nuclear crisis in Japan) have kept many militaries busy. In response to the new security landscape, many military forces had to re-organize and transform themselves to operate over a wide spectrum of operations. Figure 2-1 shows the range of operations that are highlighted in the U.S. joint operations doctrine.

FIGURE 2-1. Range of U.S. Joint Operations

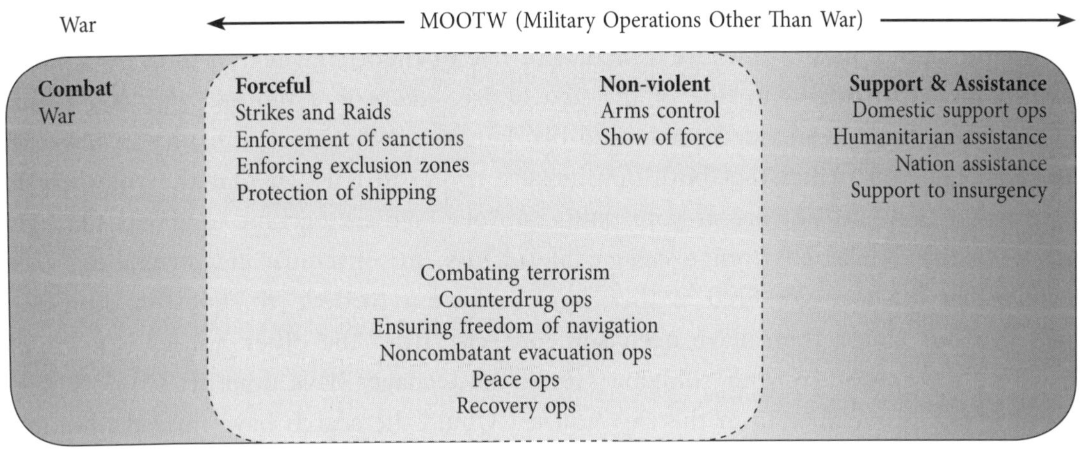

Source: Adapted from Figure 1 (p. 101) in Story, A.E. & Gottlieb, A. (1995).

Compared to the past, commanders and leaders will need to conduct a wide range of operations. Sometimes, the different types of operations may not even be conducted sequentially but simultaneously within a theater of operations. This concept, known as "Three Block War", was coined by U.S. Marine General Charles Krulak (Krulak, 1999). Conducting full scale combat, peacekeeping and humanitarian aid operations, all within the space of three city blocks is increasingly a reality today. Krulak concluded that this new condition necessitates a new breed of leaders at the lower levels, which he referred to as "strategic corporals" who are able to take independent actions and make decisions that can have strategic implications. In this new operational landscape, commanders and leaders require a greater capacity for learning and, the mental and behavioral flexibility to quickly adapt across different operations.

The Challenge of An Ill-Defined Enemy

Another great challenge facing the commanders and leaders is that the enemy in irregular warfare is no longer clearly defined by uniform, structure, weapons or tactics. The enemy does not observe the laws of the United Nations nor the established norms involving civilians in conflicts. The enemy can hide among the civilians or use them as shields and decoys to achieve their objectives. Imagine yourself as a platoon leader with your unit deployed in a peace-keeping mission in a failed state where your troops encounter insurgents and civilian sympathizers who dislike foreign interventions. While on routine foot patrol with 15 of your soldiers one day, everyone hears a shot and instinctively takes cover. No further shots are fired. There is silence and then you hear the cries of agony coming from one of your soldiers who is bleeding from a wound to his stomach. Your medic attends to him while you call for an ambulance. You discover that the shot was fired from one of the six houses across the road but no one can pinpoint the specific house. You order your soldiers to surround the houses and begin a house-to-house search with five of them. The search finds two houses with rifles but no spent cartridges. In one of the two houses, there is a mother with her teenage son while in the other, there is an elderly man and his daughter. With great difficulty in understanding the local language, you realise that both families are trying to tell you that the rifles are for self-defense and that they did not fire at your soldiers. Your soldiers become angry as they glance at their fellow soldier weakened from excessive blood loss, unconscious and struggling to survive. They believe that the teenager fired at them and seek your "orders" to "question him privately". What will you do? Could there have been someone else from the other houses who might have escaped before the cordon by your soldiers? Could the teenager have done it? Could it be one of the remaining family member from the two houses? Could the search have missed finding a third rifle among the houses?

This story illustrates how difficult it is to locate and identify an enemy in an urban warfare. The enemy can easily hide among civilians or use civilians as shields. Under such situations, commanders and soldiers are likely to face ambiguities, uncertainties, a sense of injustice and a lack of control. Ambiguity, injustice and lack of control over a situation are "stressors" that cause psychological distress which can manifest in behaviors such as anxiety, fear, frustration and anger. These behaviors are considered normal emotional reactions on the battlefield but they need to be managed in order to avoid atrocities such as the My Lai Massacre of innocent civilians (Peers, 1976). Psychological distress, if left unchecked, can lead to other behavioral problems such as self-harm. Hence, commanders and soldiers need to be even more tolerant of ambiguity and be emotionally resilient in today's operational environment.

The Challenge of Emerging Technologies

The warfare and operational landscape constantly evolve and change as emerging technologies (hardware, software and biochemical) allow military forces to create new weapons, new systems and new ways of organizing and fighting. For example, the invention and deployment of machine guns, tanks and warplanes made significant impact and changes to warfare in the first and second World Wars. In more recent wars like the U.S.-led coalition operations in Iraq and Afghanistan, the use of unmanned aerial vehicles, night vision devices, caffeine pills and network communications brought new capabilities and strategies to the battlefield, i.e., continuous intelligence gathering and operations and timely firepower against the adversary upon identification. While such technologies can provide an edge over the adversary, it can also take a toll on the commanders and soldiers. As soldiers are pushed to fight continuously and longer, they push their bodies to extreme limits and fatigue. Are commanders able to recognize when to pull back to take a break (before they lose concentration and their thinking becomes faulty) to avoid the commission of friendly fire or a breakdown in command at a critical time of operation? With greater surveillance and communications technologies, are commanders able to see the signal from the massive amount of information, make timely decisions in the face of information overload, and to pull themselves away from the digital screens and command posts to exert their command presence in the field at the critical moment?

It is without a doubt that emerging technologies will continue to drive the revolution of military affairs, and shape future warfare and operations. While it is critical for commanders to embrace the strengths of new technologies and ways of fighting, it is equally critical for them to understand the weaknesses as the enemy will certainly be looking for them. This challenge will require commanders to be vigilant and to continuously gather intelligence on the enemy's attempts to counter the new technology in order to stay ahead of the enemy.

The Challenge of Civilian Involvement

In the last two decades, civilians have increasingly become involved in wars as targets, shields and sympathizers. Each of the conditions presents a different challenge to the commander. In ethnic conflicts, civilians are the targets of ethnic cleansing in which an ethnic group is systematically killed, imprisoned, violated and/or expelled from their land by another ethnic group (e.g., Serbia-Croatia-Bosnia war in the 1990s and Rwanda genocide in 1994). In the case of ethnic cleansing and genocide, commanders in a peace-keeping operation will have to deal with the sight of the senseless violence and their sense of hopelessness or guilt when they are not able to do anything to prevent the death of civilians.

Civilians have also been used as shields as the adversary takes advantage of the rules of engagement that military forces abide by (e.g., Talebans fighting the U.S.-led coalition in Afghanistan since 2001). When civilians are used as shields, commanders will have to make difficult decisions with regard to protecting their troops versus not killing any civilians in the crossfire. It follows that they also have to cope with the guilt and repercussions when civilians are accidentally killed. Not all civilians are neutral; some may sympathize with the adversary and provide them with aid upon request. Civilians who sympathize with the adversary sow mistrust and raise the anxiety and frustration of soldiers which becomes a potential trigger for the soldiers to abuse them and treat them as the adversary. The outcomes of the three conditions point to the need for commanders and soldiers to be emotionally resilient and be anchored on professional values. From an organizational standpoint, it is also important for military forces to have clear rules of engagement, procedures and resources for psychological debriefing and support for their soldiers when they encounter civilian casualties.

The Challenge of Media Everywhere

The Vietnam war has been widely regarded as the "living room war". Journalists brought images of burning children and American casualties to the television screens nationwide, and newspapers reported equally horrifying scenes. The American war coverage has since been viewed by many in

A group of media personnel following Israeli soldiers during an operation in West Bank in 2002.
Source: Scott Nelson/Getty Images News/Getty Images

the military as responsible for undermining the war effort, while others called it the turning point in the home support for the war. Hence, the view that the Vietnam war was fought and lost in the living rooms of America. Today's wars are fought not only in the living room and newspapers but over the Internet and handphone as well. Advances in satellite and communications technologies have the capability to allow video and still images to be broadcast instantaneously from the war zone to any media platform. Furthermore, advances and the availability of recording devices on cameras and handphones have allowed any civilian to be a journalist, to capture relatively clear footage of any commander in action and put it on the World Wide Web for all to see. Soldiers themselves carry these devices and record their own actions for fun. Some of these inadvertently get passed on or seized and make their way to the media as in the infamous incident of prisoners' abuse in Abu Ghraib prison, Iraq, in 2004 by U.S. soldiers.

Many military forces today provide training and have guidelines for commanders and men to interact with journalists and the media. Commanders ought to know and apply the guidelines accordingly. There is little a commander can do to prevent or control civilians from taking pictures or recording their actions. In the final analysis, commanders should have nothing to fear if they do their jobs professionally and ethically, and train their men to carry out their duties equally well.

CONDITIONS AND STRESSORS IN OPERATIONS AND BATTLEFIELD

As highlighted earlier, there are many types of operations and the one that is most stressful and violent would be the one involving combat. In traditional warfare, combat takes place in the battlefield. Today, combat is more likely to be part of an ongoing operation wherever it may be. Regardless of whether combat takes place in the battlefield or in today's wide spectrum of operations, war veterans, historians and researchers have found that soldiers face similar conditions and stressors, and exhibit similar reactions when intense fighting takes place. Some of them are well described in classic books such as *Men Against Fire* (Marshall, 1947), *Acts of war: the behavior of men in battle* (Holmes, 1986) and *Soldiers: A history of men in battle* (Keegan, Holmes & Gau, 1985). The list of conditions and stressors (i.e., stimuli in the environment that potentially cause stress) is drawn from these readings and other more recent scholarly writings and military research (e.g., Adler, Castro & Britt, 2006; Campbell & Nobel, 2009; Hancock & Szalma, 2008; and NATO RTO, 2008).

Harsh Physical Environment and Working Conditions

All military forces train their soldiers to operate in tough physical environment and conditions to prepare them for operations and war. Nevertheless, often for safety reasons, training falls short of the extreme conditions and demands placed on the soldiers in operations. Krueger (2003, 2008) provides a comprehensive description of the physical conditions and impact of such conditions on soldiers' performance. These conditions include extreme environmental temperatures, high altitude, noise, acceleration, vibration, unnatural motion, and toxic gases and fumes. A soldier typically spends much of his/her time dealing with the harsh environmental and working conditions and only a relatively short moment fighting the enemy. Commanders should not underestimate the effects in which these elements drain the energy and affect the mental capacities (e.g., attention span and psycho-motor skills) and motivation of the soldiers to perform.

Deprivation and Fatigue

The harsh environment, demands of mission and modern technologies that facilitate continuous operations can quickly cause sleep deprivation and fatigue in soldiers. Studies have shown that one day of sleep loss can result in the deterioration of vigilance and mental functions (e.g., decision making) and three days of sleep loss can lead to severe impairment and even hallucinations and paranoia. Fatigue will be made worse if soldiers are also deprived of food, water and basic hygiene when logistic supplies are disrupted or delayed. Sleep management and basic provisions are essential for soldiers to function effectively. To this extent, the Australian Defence Force Psychology Organization and the Defence Science and Technology Organization have produced a commander's guide (see Australian Government Department of Defence, 2002) on "Fatigue Management During Operations", detailing useful information on sleep, signs of fatigue, effects of sleep loss and fatigue on performance, and techniques to prevent and manage fatigue. The guide (p. 69) indicates that "As a guide, general combat effectiveness can be maintained on three hours of sleep per 24-hour period. Most soldiers can maintain an acceptable level of physical performance for 9-14 days on such a schedule. Tasks that are mainly mental in nature require more sleep." It went on to provide recommendations (see Table 2-1 below)

TABLE 2-1. Minimum Operational Sleep Requirement

Mainly Physical Tasks			*Mainly Mental Tasks*		
Acceptable performance (days)	**Sleep required (uninterrupted)**	**Suggested recovery period**	**Acceptable performance (days)**	**Sleep required (uninterrupted)**	**Suggested recovery period**
3	0	24 hrs	1	0	8–12 hrs
6	1.5 hrs per day	2 days	2	1.5 hrs per day	12–24 hrs
9	3 hrs per day	3 days	3	3 hrs per day	24 hrs

Source: Reprinted with permission from Australian Government Department of Defence.

which is accredited to British Army research on the minimum amount of sleep required to maintain acceptable performance for sustained operations.

In an overseas deployment, other deprivations would include lack of contact with loved ones, lack of family support, lack of privacy, lack of leisure activities and material comforts that one is used to having back home. In addition to these, soldiers may worry about the fidelity of their girlfriends and wives, the proper upbringing of their children and the care of their aging parents. Commanders should not underestimate the effects of these depivations on the mental capacities, health and motivation of the soldiers to perform.

Enemy threat

The enemy is a main stressor because it poses a direct threat to the soldier's life. Soldiers will be concerned about whether the adversary is more skillful, well-equipped, fitter, motivated, intelligent, aggressive and ruthless than them. The image of the enemy may not be homogenous — it is possible for soldiers to see different strengths and weaknesses of the enemy as an individual, a unit/regiment and as a military force. Such perceptions of the enemy are likely formed from past knowledge and experiences, and updated by recent encounters, rumours and information warfare. Soldiers will often not voice their thoughts and fears about the enemy for fear of reprisals and ridicule. However, it is important that commanders identify and address such fears early as they may otherwise lead to flight or surrender on the battlefield. On the other hand, commanders should also be concerned when soldiers undersestimate the enemy's ability and show arrogance and contempt for the enemy.

Fog of War

This is a common term used by the military to refer to the confusion and disorientation in the battlefield created by limited visibility, noise, ambiguous and conflicting information, and surprises and unexpected actions by the enemy and own forces. Indeed, this is the nature of combat in which each side attempts to disorient and surprise the other side using all means so as to gain victory. Stressors such as uncertainty, ambiguity and surprises are not only confined to the battlefield but appear in almost every part of an operation.

Destruction and Casualties

The nature of war involves destruction and casualties. For soldiers who experience the sight, smell and sound of war for the first time, these experiences can be very stressful and traumatic. Encounters with the horrors of war are likely to lead soldiers, at some point of their deployment, to question and re-examine their beliefs about humantiy, morality, the purpose of their mission, their role and self-identity. Such close personal encounters and experiences can have a life changing impact on a soldier's mental health and functioning.

THE STRESS PROCESS AND OUTCOMES

A Model of Stress

There are many excellent publications and writings on stress and combat/operational stress. The following are some that have contributed to the content in this chapter: Alder, Castro and Britt's (2006) book on operational stress; Bartone's (2003) book chapter on the different stressors in a number of U.S. Army operations; Driskell and Salas' (1996) book on stress and performance; Grossman's (2004) book on the physiological and psychological effects of combat and killing; Hancock and Szalma's (2008) book on performance under stress, Soloman's (1993) book on combat stress reaction; U.S. Army's (2000, 2009) manuals and U.S. Marine Corps' (2007) guide on combat stess; and NATO RTO's (2008) report on stress and guide on psychological support.

There are also many definitions and conceptions of stress among the scholars (e.g., Aldwin, 2007; Campbell & Nobel, 2009; Jones & Bright, 2001; Lazarus & Folkman, 1984). The following proposed

FIGURE 2-2. A model of stress in the military

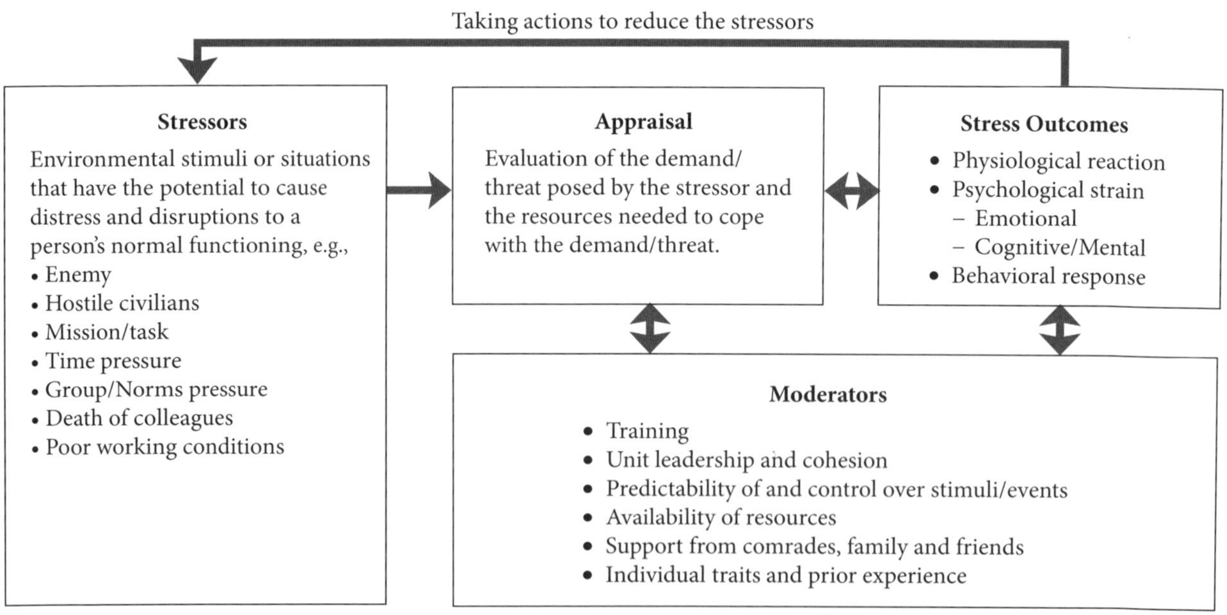

model (see Figure 2-2) is not meant to be comprehensive but to serve as an illustration based on a common understanding of stress and its relevance to the military.

Definition of stress. The use of the word "stress" is commonly associated with a negative emotional or mental state (e.g., anxiety, worry, fear) accompanied by some physiological and behavioral symptoms (e.g., increased heart rate, perspiration agitation, impatience, insomnia). But to many scholars, what is commonly known as "stress" is an outcome that is better labeled as "distress" or "strain". Stress is best viewed as a process that occurs when the person tries to deal with the stressors in the environment.

The Stress Process

In the stress process, the individual consciously or automatically appraises the demand or threat of the stimuli/situation and the resources available to cope with it (this may be his/her own ability or what is available to him/her to use), which in turn produce the stress outcomes. The intensity and nature of the stress outcomes depend on the perceived gap between the demand/threat and coping ability, the influence of the moderators and the nature of the stressor. For example, if the soldier perceives the enemy as more aggressive and well-equipped than his/her own unit, the soldier will be stressed when he/she thinks about fighting the enemy. The stress is in fact anxiety, or fear when it is more intense and specifically linked to a stimulus or outcome. A moderator of fear would be whether the soldier feels he/she has been trained well to exploit the enemy's weaknesses using his/her weapons and equipment, or whether he/she has confidence in his/her commander in outwitting the enemy on the battlefield.

Fear. Anxiety and fear are unquestionably the most common and intense of emotions experienced by soldiers in operations and on the battlefield. Afterall, the central issue in combat is about "kill or be killed". Besides the fear of dying, there is the fear of being maimed, of being captured, of failure, of letting one's comrades or unit down, of the unknown, of surprises and so on. Fear is, therefore, normal in operations and on the battlefield. It is uncontrolled fear which can be contagious and can lead to paralysis or flight that is detrimental to performance and mission success.

Stress Response: Fight, Freeze or Flight (and More . . .)

Stress (or fear) arouses the body and prepares it for fight or flight. So, a certain amount of stress is actually ideal for combat. When the body is aroused by the nervous system, certain hormones are released triggering heart rate increases, redistribution of blood to vital organs and muscles, constriction of minor blood vessels and the dilation of pupils. Grossman (2004) (see also Grossman & Siddle,

2000) suggests that the ideal heart rate should be about 115 beats per minute (bpm) for optimal fine motor skills, and between 115 and 145 bpm for optimal complex motor skills, visual reaction time and cognitive reaction time. When the heart rate is beyond the ideal zone (which will vary among individuals), the soldier starts to lose cognitive processing ability, peripheral vision, depth perception, near vision and hearing. At the level of 175 bpm and above, the soldier loses control — the soldier may freeze, flee or charge irrationally at the enemy, and may involuntarily urinate and defecate. The intensity of the physiological, biological and psychological response of the body in near death situations or the death of a fellow soldier in close combat can traumatize the soldier into developing combat stress reaction, and for some, post-traumatic stress disorder or PTSD, which is a long term effect of trauma. PTSD is a medically classified term.

Combat Stress Reaction and Casualties

The psychological breakdown on the battlefield which is often temporal is now commonly known as "Combat Stress Reaction" (CSR). Previously, CSR was labeled as "shellshock", "battleshock" and "battle fatigue". Combat stress casualties refer to soldiers who cannot function as combatants not because of physical injuries but because of CSR that produces strange behaviors that endanger the soldiers themselves and others (Noy, Gal, & Mangelsdorff, 1991; Soloman, 1993; U.S. Army, 2000 & 2009). CSR does not have a formal psychiatric defintion yet mainly because it has such a diverse range of manifestations not only across soldiers but even within soldiers across time. The U.S. Army (2000, p. 2) field manual describes CSR as a collection of adverse behaviors, as a function of exposure to danger, that "produce disruptive behaviors and threaten individual and unit safety". Some of the more severe behaviors/symptoms include trembling, ducking at sudden sounds and movement, immobility of hand, arm or leg for no apparent physical reason, staggering and swaying when walking or standing, hysterical outbursts, crying, stuttering, talking inappropriately, inability to speak, sight or sound of things that do not exist, memory loss, apathy and indifference to danger.

The incidence of CSR is known to be related to the intensity and duration of close combat. In

Stress — a silent and dangerous "enemy".
Source: iStockphoto (left) and Joe Raedle/Reportage/Getty Images (right)

BOX 2-1

Mental Health of U.S. soldiers in Operation Iraqi Freedom (OIF) in 2009

The following statistics were extracted from the U.S. Army Medical Command Mental Health Advisory Team (MHAT) VI (8 May 2009) report which is based on surveys of over 2000 soldiers in Iraq from December 2008 to March 2009.

- Mental health problems (which are defined as acute stress, depression and anxiety) in maneuver units are 11.9% and in support and sustainment units, 12.3%.
- Divorce or separation intent is 16.5% in maneuver units, and in support and sustainment units, 17.2%.
- Since the beginning of OIF, there have been 162 confirmed suicides in the Iraqi Theater of Operations. There were 34 in-theater suicides in 2008 which is equivalent to an annual rate of 21.5 per 100,000 U.S. Service Members.
- Lower morale and more mental health problems were reported by soldiers on their second and third/fourth deployment. It is worse for support and sustainment than maneuver units.

fierce battles, the ratio could be as high as 1:1 combat stress casualty for every wounded-in-action (WIA). During World War II in Europe from 1942 to 1945, it was estimated that the ratio was 1:3 (U.S. Army, 2000).

The sum of all stressors. As highlighted earlier, there are many stressors in an overseas operation such as home problems, poor daily working conditions, constant threat from the enemy and close combat. When the soldiers cannot cope with all the stressors, they may manifest serious emotional, mental and behavioral problems (see Box 2-1) which will require professional help.

STRESS MANAGEMENT AND BUILDING RESILIENCE

A command responsibility. It is the commander's responsibility to manage the stressors in the environment for the soldiers as best as he/she can, educate soldiers on how to recognize stress symptoms and use positive techniques of coping, create a unit climate in which soldiers watch out and support each other, provide resources (e.g., rest and time) for soldiers to cope, and be proactive in identifying soldiers who may need immediate counseling by the commander or referral to mental health professionals. To this end, the U.S. Army in 2009 produced the field manual *Combat and operational stress control manual for leaders and soldiers* and NATO Research and Technology

Organization had produced *A leader's guide to psychological support across deployment* (NATO RTO, 2008, Annex G) which is the result of a four-year project involving commanders and professionals across various NATO countries.

Stress Management Techniques

There are many guides on stress management techniques or coping strategies. Some of the common recommendations are highlighted below. Commanders can use them and share them with their soldiers:

- Do deep breathing exercise to calm down quickly (lower their heart rates)
- Focus on what you can do, make plans and take steps to deal with (alter or avoid) the stressor
- Look for the silver lining or opportunities in the situation; look at the bigger picture
- Re-examine your beliefs and attitude toward the stressor
- Talk to someone about the stressor; do not bottle up your emotions
- Get out and do some exercises
- Get some rest and do the things you enjoy and find relaxing
- Do not drink, self-medicate or smoke to numb your emotions/nerves

Hardiness and Battlemind

The beneficial effects of realistic training on fear and stress are undisputed (e.g., Horn, 2004; NATO RTO, 2008). Good training is essentially a stress inoculation exercise in which soldiers are put through many similar stressors that they will face in operations. Through training, soldiers become fit and learn to cope with different stressors individually as well as a unit in a fairly safe and protected environment, thus increasing their self-confidence and resilience to the harsh environmental elements and deprivation in operation.

As there is no real destruction or threat of being killed in training, soldiers should be educated on what to expect on the battlefield and in close combat. Awareness of such key information is important as it reduces fear of the unknown and surprises when they happen which would otherwise psychologically dislocate them at the critical moment. Soldiers should be educated on the body's physiological, psychological and behavioral response to extreme stress, optimal zone for functioning, signs of CSR, that fear and CSR are normal reactions to threats and an extremely stressful situation respectively, and how they will be managed if they develop CSR. Utilize the buddy system to emphasize the expectation and obligation to watch out for stress symptoms in one another and to take necessary actions to address or report them.

Much of the education should be incorporated into exercises and integrated across training environments (Thompson & McCreary, 2006) where soldiers get to recall and apply their knowledge regularly so that they develop discipline, instinctive behaviors and routines that are important for their survival and managing their fear in combat. Related to this is the notion of hardiness personality (Kobasa, 1979) which develops early in life, is fairly stable but still amenable to change. Bartone (2006) described hardy persons as having "a high sense of life and work commitment . . . open to change and challenges . . . interpret stressful and painful experiences as a normal aspect of existence . . ." (p. 137). He reported that hardiness moderated the level of combat stress of soldiers during the Gulf War, and put forth the argument that hardy leaders develop hardy soldiers by influencing their thinking and behavior.

Good leaders provide realistic training, build unit cohesion, and act and communicate in ways that inspire confidence and reduce fear in their soldiers. Good leaders come from selection, training and education. Good training and education includes updating and reminding leaders what to look out for and what to do during operations (see Box 2-2 on building "Battlemind"). In this way, both the hardiness of the leaders and soldiers may be systematically and consistently strengthened over time to act as a vaccine against combat stress.

Besides building the soldiers' resilience to the combat environment, commanders must also pay attention to their soldiers' family and home support. Soldiers today are more likely to be deployed for overseas missions and repeatedly, family or home problems can be highly distracting and stressful. NATO RTO (2008, Annex G) leader's guide highlighted this need and provided specific recommendations for each phase of deployment. For example, during pre-deployment, leaders should educate their subordinates on what to expect for the deployment and family adjustment, hold pre-deployment briefings for family members, and establish a rear detachment/family support group setup. During initial deployment, leaders can send regular informal updates to family members and support rear detachment/family support group activities that help the families cope. There is also much to be done at the post-deployment stage to prepare the soldiers and their family members as there are possible re-integration issues — soldiers' outlook might have been affected by events during deployment and their family members would have been accustomed to routines without the role of the husband/father.

Combat Stress Reaction Management

When soldiers develop CSRs, commanders are likely to be the first responders. Many CSRs are temporal and do not require medical treatment. Soldiers affected by CSRs should be treated in the unit as near the front as practical and possible, and not be evacuated to the rear hospital as medical casualties (a strategy known as "Proximity"). This is because research has shown that they have a higher chance of recovery when treated at the front than at the rear. Research has also found that most CSR soldiers simply need rest, food and sleep and these should be provided as soon as possible for their recovery

BOX 2-2

Battlemind

The following content was extracted from the brochure titled "10 tough facts about combat: What leaders can do to mitigate risk and build resilience", produced by the Walter Reed Army Institute of Research Land Combat Study Team (2006). It was intended for commanders in Operation Iraqi Freedom. An illustration is provided for "Fact #1"; please see the original document for the "findings" and "what leaders can do" for the remaining nine facts.

"Battlemind" is a soldier's "inner strength to face adversity, fear, and hardship during combat with confidence and resolution. It is the will to persevere and win".

Fact #1: Fear in combat is common.

Findings:

- Over 2/3 of silver star recipients reported increased fear as battle progressed.
- Common symptoms of fear: violent shaking/trembling, losing control of bowels, feeling weak, cold sweats, and vomiting.
- Fear and anxiety are reduced in combat when soldiers engage in actions derived from their training experiences.

What leaders can do:

- Drill and train soldiers in specific actions to take under combat conditions. Tough training is the best preparation.
- Provide soldiers sufficient physical and mental reset time.
- Admitting and joking about fear will release tension.
- Remember that fear is NOT a mental disorder.

EVEN HEROES FEEL FEAR.

Fact #2: Unit members will be injured and killed.

Fact #3: There will be communication and information breakdown.

Fact #4: Soldiers frequently perceive failures in leadership.

Fact #5: Combat impacts every soldier mentally and emotionally.

Fact #6: Combat often leads to lasting adverse mental health effects.

Fact #7: Soldiers are afraid to admit that they have a mental health problem.

Fact #8: Deployments place a tremendous strain upon families.

Fact #9: The combat environment is harsh and demanding.

Fact #10: Combat poses moral and ethical challenges.

to begin (a strategy known as "Immediacy"). Lastly, it is important for commanders to reassure the CSR soldiers that what they are experiencing is a normal reaction to extreme stress (that they are not weak and they are not casualties) and they will recover and return to full duty in a few hours or days (a strategy known as "Expectancy"). Taken together, the acronym **PIE** can help commanders to remember what to do in such situations (Noy, Gal, & Mangelsdorff, 1991; Soloman, 1993).

The U.S. Army (2009) recommends **BICEPS**, which is an extended and elaborated version of the same principles. **B** is for "Brevity" which emphasizes the short treatment of 1 to 3 days of rest and replenishment in the unit. **I** is for "Immediacy". **C** is for "Contact" which emphasizes the expectancy of functioning as a soldier again facilitated by maintaining contact with the unit and the soldier's leader during his/her recovery. **E** is for "Expectancy". **P** is for "Proximity". **S** is for "Simplicity" which emphasizes the use of brief and straightforward methods to restore the soldier's physical well-being and self-confidence.

"If treated near their units, 65 to 85 percent of combat stress casualties . . . return to duty within 1 to 3 days. About 15 to 20 percent more return to duty in 1 to 2 weeks. Only 5 to 10 percent are sent home, and these usually have other problems in addition to combat stress reactions" (U.S. Army, 2000, p. 54). If the symptoms do persist, the soldiers will have to be examined by the medical or mental health professionals.

Traumatic Event Management

Not all combat situations or operations produce CSR in soldiers. What should a commander do after an operation or accident that is potentially a traumatic experience for many of his/her soldiers? An event is "potentially traumatic when it causes individuals or groups to experience intense feelings of terror, horror, helplessness and/or hopelessness. Guilt, anger, sadness, and dislocation of world view or faith are potential emotional/cognitive responses" (U.S. Army, 2009, p. 2–21). As such events can potentially affect the soldiers' psychological health, self-confidence and working relationship with one another, it is important for leaders to consider conducting restorative debriefing sessions (with the help of mental health professionals) so that the individuals' and team's functioning would not be degraded. When trained professionals are not available, the U.S. Army recommends "cool-down meetings" and "leader-led after-action debriefing".

A cool-down meeting (U.S. Army, 2009) is a short informal meeting (about 15 minutes) that is held at a safe and relatively comfortable location immediately after the potentially traumatic event, before any formal debriefing or rest. The time is used for the leader to calm the team by complimenting any positive aspects of the event, providing information on what to expect next (including any debriefings or mission), and providing an opportunity for the team to talk among themselves and consume some liquids and convenience food (if available).

According to the manual (U.S. Army, 2009, p. 2–23), the leader-led after-action debriefing (LLAAD) "should be conducted after all missions especially when the maneuvers did not go according to plan" and when there are injuries and death. It is appropriate only at the platoon, squad or team level. The team may be the battalion commander and his/her immediate staff, or any other similar configuration. The best time to conduct the debriefing is after the members have replenished bodily needs and are no longer in a highly aroused state. The general guide is that the leader should not run it alone if there are trained professionals to assist or consult. This is because such debriefings not only reconstruct the facts of the event but also allow all members to voice their thoughts, reactions and emotions which may evoke sadness, guilt, anger and blame that can seriously threaten the unit's/team's cohesion or arouse dysfunctional emotions and behavior in individuals if they are not handled properly. The procedure is generally similar to Critical Incident Stress Debriefing (Mitchell & Everly, 2001) in which mental health professionals are trained to facilitate and followup with individuals who are badly affected by the traumatic experience.

Behavioral Science and Mental Health Professionals

Military organizations all over the world recognize the challenges and stressors that soldiers face and the leaders will have a great impact on the soldiers' performance, well-being and health. As such, many have established institutions that employ behavioral science and mental health professionals and scientists (e.g., psychologists, sociologists, social workers and psychiatrists) to provide direct helping services (e.g., mental health assessment, counseling and treatment), education and training (e.g., team-building and stress management), advice on policies, doctrines and systems development (e.g., organizational climate survey-feedback, doctrine for leadership development and selection system for commanders), and research (e.g., how to build swift trust in an ad hoc team and how to enhance soldiers' resiliency). Many of these military professionals and researchers meet annually at conferences such as the International Military Testing Association Conference (www.internationalmta.org) and the Society of Military Psychology (http://www.apa.org/division/div19/about/html) Symposium. It is important for all commanders to be aware of such institutions and professionals in their own military forces so that they know where to seek help for their soldiers and to draw on the resources available to enhance their own leadership and unit effectiveness.

SUMMARY

A soldier's job is to learn from past battles, and anticipate and prepare for future wars so as to achieve victory over the adversary. This chapter's aim was to show and contrast the "new" and the "old" challenges and stressors in operations. Unfortunately, the "new" does not replace the "old" but adds to and complicates it. The "new" being the challenges of the modern operation such as multiple and complex security threats, ill-defined enemy, emerging technologies, civilian involvement, and media everywhere. The "old" (such as harsh physical environment and work conditions, deprivation and fatigue, enemy threat, fog of war, destruction and casualties) being the conditions and stressors in conventional warfare. A model of stress was presented (Figure 2-2) to help understand soldiers' response to stressors and the basis of fear.

Fear is a common response in battle but it needs to be controlled. Extreme fear and stress negatively affect soldiers' performance and can even lead to CSRs which, if not managed well, could lead soldiers to endanger themselves and others. It is a commander's responsibility to help soldiers manage their stress and there are many positive actions that commanders can take. Some basic applications such as stress management techniques, building resilience and strategies for management of combat stress reaction are highlighted. Much of the knowledge comes from the behavioral sciences and the mental health profession as well as from researchers who work in military forces. Commanders can rely on such professional resources in their own military forces and seek professional assistance to enhance their soldiers' performance and their units' combat readiness and effectiveness.

Key Concepts

- Military Operations Other Than War (MOOTW)
- Irregular warfare
- Fog of war
- Fear
- Stress
- Stressors
- Stress outcomes
- Stress response
- Stress management
- Post-Traumatic Stress Disorder (PTSD)
- Combat Stress Reaction (CSR)
- Resilience
- Hardiness
- Battlemind
- Proximity, Immediacy, Expectancy (PIE)
- Brevity, Immediacy, Contact, Expectancy, Proximity & Simplicity (BICEPS)
- Traumatic event management

Questions

- As a military leader, what type of operations are you likely to be involved in?
- Based on what you have read in this chapter, identify the challenges and stressors you and your soldiers will face in the operations.
- To what extent are you and your soldiers prepared and trained for such operations? Elaborate on the training that you and your soldiers have received to help you manage and overcome the stressors in operations.

References

Adler, A.B., Castro, C.A., & Britt, T.W. (Eds.). (2006). *Military life: The psychology of serving in peace and combat (Vol. 2): Operational stress.* Westport, CT: Praeger Security International.

Aldwin, C.M. (2007). *Stress, coping, and development: An integrative perspective.* New York: Guilford Press.

Australian Department of Defence (2002). *Fatigue management during operations: A commander's guide.* Puckapunyal, VIC: Land Warfare Development Centre.

Bartone, P.T. (2003). Stress in the military setting. In C. Cronin (Ed.). *Military psychology: An introduction* (pp. 113–148). Boston, MA: Pearson.

Bartone, P.T. (2006). Resilience under military operational stress: Can leaders influence hardiness? *Military Psychology, 18*, 131–148.

Campbell, D.J., & Nobel, O.B.-Y. (2009). Occupational stressors in military service: A review and framework. *Military Psychology, 21* (Suppl 2), 47–67.

Deci, E.L., & Ryan, R.M. (1985). *Intrinsic motivation and self-determination in human behavior.* New York: Plenum.

Deci, E.L. & Ryan, R.M. (2000). The "what" and "why" of goal pursuits: human needs and the self-determination of behavior. *Psychological Inquiry, 11*, pp. 227–268.

Driskell, J.E., & Salas, E. (1996) (Eds.). *Stress and human performance.* Mahwah, NJ: Lawrence Erlbaum Associates.

Grossman, D. (2004). *On combat: The psychology and physiology of deadly conflict in war and in peace.* PPCT research publications.

Grossman, D., & Siddle, B. (Mar 2000). *Psychological effects of combat.* Retrieved Jan 2, 2011, from http://www.killology.com/article_psychological.htm.

Hancock, P.A., & Szalma, J.L. (Eds.). (2008). *Performance under stress.* Burlington, VT: Ashgate Publishing.

Hedges, C., & Al-Arian, L. (2008). Collateral damage: America's war against Iraq civilians. New York: Nation Books.

Holmes, R. (1986). *Acts of war: the behavior of men in battle.* New York: Free Press.

Horn, B. (2004). The worm revisited: An examination of fear and courage in combat. *Canadian Military Journal, 15*, 5–16.

Jordan, D., Kiras, J.D., Lonsdale, D.J., Speller, I., Tuck, C., & Walton, C.D. (2008). Understanding modern warfare. Cambridge, U.K.: Cambridge University Press.

Jones, F., & Bright, J. (2001). *Stress: Myth, theory and research.* Harlow, U.K.: Prentice Hall.

Keegan, J., Holmes. R., & Gau, J. (1985). *Soldiers: A history of men in battle.* London: H. Hamilton.

Kobasa, S.C. (1979). Stressful life events, personality and health: An inquiry into hardiness. *Journal of Personality and Social Psychology, 37*, 1–11.

Krueger, G.P. (2003). Military performance under adverse conditions. In C. Cronin (Ed.). *Military psychology: An introduction* (pp. 89–112). Boston: Pearson.

Krueger, G.P. (2008). Contemporary and future battlefields: Soldier stresses and performance. In P.A. Hancock & J.L. Szalma (Eds.), *Performance under stress.* Burlington, VT: Ashgate Publishing.

Krulak, C.C. (1999). The strategic corporal: Leadership in the three block war. *Marine Corps Gazette, 83(1)*, 18–23.

Lazarus, R.S., & Folkman, S. (1984). *Stress, appraisal, and coping.* New York: Springer Publishing Company.

Marshall, S.L.A. (1947). *Men against fire.* New York: William Morrow.

Mitchell, J.T., & Everly, G.S. Jr. (2001). *Critical incident stress debriefing: An operations manual for CISD, defusing and other group crisis intervention services* (3rd ed.). Ellicott City, MD: Chevron Publishing Corporation.

NATO Research and Technology Organization (2008). *Stress and psychological support in modern military operations* (RTO Technical Report, TR-HFM-081). Retrieved Jan 2, 2011, from http://www.rta.nato.int/Pubs/RDP.asp?RDP=RTO-TR-HFM-081.

Noy, S., Gal, R., & Mangelsdorff, A. D. (1991). Combat stress reactions. In *Handbook of military psychology.* (pp. 507–530). Oxford England: John Wiley & Sons.

Peers, W.R. (1979). *The My Lai Inquiry.* New York: W.W. Norton and Co.

Soloman, Z. (1993). *Combat stress reaction: The enduring toll of war.* New York: Plenum Press.

Story, A.E., & Gottlieb, A. (1995). Beyond the range of military operations. *Joint Forces Quarterly, 9,* 99–104.

Thompson, M.M., & McCreary, (2006). Enhancing mental readiness in military personnel. In A.B. Adler, C.A. Castro & T.W. Britt (Eds.), *Military life: The psychology of serving in peace and combat (Vol. 2): Operational Stress.* (pp. 54–79). Westport, CT: Praeger Security International.

Thussu, D.K., & Freedman, D. (2003) (Eds.). *War and the media: Reporting conflict 24/7.* Thousand Oaks, CA: Sage.

U.S. Army (2000). *Field Manual 6-22.5: Combat stress.* Washington, D.C.: Headquarters, Department of the Army.

U.S. Army (2009). *Field Manual 6-22.5: Combat and operational stress control manual for leaders and soldiers.* Washington, D.C.: Headquarters, Department of the Army.

U.S. Army Medical Command Mental Health Advisory Team (MHAT) VI (May 2009) *Operation Iraqi Freedom 07–09.* Retrieved Jan 2, 2011, from http://www.armymedicine.army.mil/reports/mhat/mhat_vi/MHAT_VI-OIF_Redacted.pdf.

U.S. Department of Defense (2006). *Joint publications 3-0: Joint Operations.* Washington, D.C.: Department of Defense.

U.S. Marine Corps (2007). *Combat stress: A concept for dealing with the human dimension of urban conflict.* Arlington, VA: Headquarters, Marine Corps.

Van Creveld, M. (2008). *The changing face of war: Combat from Marne to Iraq.* New York: Presidio Press.

Walter Reed Army Institute of Research Land Combat Study Team (2006). *10 tough facts about combat and what leaders can do to mitigate risk and build resilience.* Retrieved Jan 2, 2011, from http://www.armyg1.army.mil/dcs/docs/10%20Leaders%20Tough%20Facts%20About%20Combat%20Brochure%2011%20SEP%2006.pdf.

3

SOLDIER PERFORMANCE: The Psychology of Human Behavior in Combat

Source: CyberPioneer, Ministry of Defence, Singapore

OVERVIEW

What comes to mind when you think of a soldier's combat performance? For many of us, we may think of an infantry soldier bravely charging at the enemy despite their suppressive fire and lobbing a grenade to take them out. Those who are in the Air Force may think of a fighter pilot doing 8Gs, outmaneuvering and destroying an enemy aircraft with a single missile. For those in the Navy, it could the image of a Commanding Officer calmly giving out a string of orders to launch a missile to destroy an enemy ship while at the same time defending itself against enemy attacks from the air. While these actions clearly demonstrate combat performance, what about others who deliver their combat missions but get killed or seriously wounded during the fight or carry out actions that accidentally kill their fellow soldiers? What about other actions that involve mistakes that have serious consequences for their own forces?

As soldier performance determines mission success, and commanders are responsible for both mission success and the behaviors of their soldiers, it is important for commanders to have a good understanding of what constitutes performance and the factors that promote positive or negative performance. This knowledge will help commanders to anticipate as well as recognize these factors when they occur so that they are primed to think of actions to facilitate positive performance and prevent negative performance. This chapter will introduce a framework to guide thinking and conceptualisiing of soldier performance. Soldier performance is essentially composed of a set of behaviors that helps to achieve the mission or objectives of the military. Hence, knowledge of the psychology of human behavior is essential for the understanding of the factors that influence performance. This chapter on the psychology of human behavior therefore sets the foundation for understanding performance as well as the topic of stress reaction in the previous chapter and the content in the next few chapters (Chapters 4, 5 and 6) of this book.

SOLDIERS' PERFORMANCE

Much has been written on military performance under stress (e.g., Britt, Castro & Adler, 2006; Driskell & Salas, 1996; and Hancock & Szalma, 2008), and especially on factors affecting soldiers' performance. But what is performance? Are commanders clear about what constitutes performance from their soldiers? Industrial-organizational psychologists have studied "performance" and their research provides some interesting insights into understanding what constitutes performance. Performance is "something that people actually do and can be observed . . . (and) includes only those actions or behaviors that are relevant to the organization's goals . . . " (Campbell, McCloy, Oppler & Sager, 1993, p. 40). Borman and Motowidlo (1993, 1997) further refined the concept of performance to comprise *task performance* and *contextual performance* which are both important for organizational effectiveness.

Task Performance

Task performance is about doing activities and achieving the tasks and goals that are prescribed by the role and job (or mission in the military). Task performance bears a direct relation to the organization's core purpose and competence. For example, the infantry section leader is expected to be proficient in his/her camouflage, navigation, execution of section-level drills and shooting. A pilot, is expected to be proficient in flying and destroying enemy targets with the missiles/bombs. The sum of all task performances contributes directly to winning a war. For example, these task performances could include the number of territories captured or enemy kills achieved. Such measurable outcomes can then be used to reflect the effectiveness of a military organization in achieving its purpose and competence.

Researchers have also found that task performance varies in intensity of effort and achievement. Performance is commonly classified as *maximum* and *typical* performance (e.g., Sackett, Zedeck and Fogli, 1988). Soldiers demonstrate maximum performance when they give their best effort in pursuit of a desired outcome, for example, excelling to top a test or pushing their bodies to extreme endurance or strength in a combat situation. Typical performance is commonly observed in daily activities where effort is expended to sufficiently meet the requirements of the task. It is quite common to find that soldiers in training often retain some energy in anticipation of the next task to be performed.

As one reads further into the subject of performance, one finds that there is what could be considered as sub-standard or *poor performance*, followed by *non-performance* or no action taken, and finally *negative task performance* in which wrong actions (e.g., mistakes or errors) are committed that result in opposite effects of the intended tasks. Sub-standard performance in soldiers can be a result of poor training where soldiers may not have acquired the proficiency levels required to perform specific tasks. It could also be a result of a lack of motivation to carry out the task even if the soldiers meet the knowledge and skill requirements. In some cases soldiers may take no action (i.e., non-performance) because they are either completely confused about the task requirements or they deliberately choose not to take any action to signal dissatisfaction or defiance. It could also be because they are mentally incapacitated by the extremely stressful situation (see "combat stress reaction" in Chapter 2).

Finally, soldiers also make mistakes or take actions that are contrary to the behavior required of task performance. Such negative task performance derails mission and operational objectives, and may hurt the morale and effectiveness of the unit and their military organization. The accidental shooting of civilian aircrafts and friendly fire are examples of unintentional mistakes and inappropriate actions of soldiers in executing their tasks. Such mistakes may lead to severe negative consequences. In addition to mistakes, negative task performance also includes behavior which is not only wrong but deliberately carried out, for example, the abusive and derogatory actions of U.S. soldiers toward the prisoners under their charge in the Abu Ghraib incident (Carter, 2004). When exposed, such incidents could seriously affect the credibility of the military organization and home support for the operation.

Contextual Performance

Contextual performance refers to intentional discretionary behavior that is not prescribed by the role, job or task. It is closely related to Organizational Citizenship Behavior (Organ, 1997). For

Helping a fellow soldier without being asked or instructed, especially one who is from outside the unit, is an example of contextual performance.
Source: Joe Raedle/Edit/Getty Images

simplicity, the term "contextual performance" is used in this book. Contextual performance (Borman & Motowidlo, 1997) is about behavior (person-directed) that helps other organizational members perform in their jobs and/or behavior (organization-directed) that helps support the attainment of organizational goals. These include offering to help others in the organization without them asking, volunteering for extra assignments, making suggestions to improve the organization and presenting the organization favorably to outsiders. In military jargon, such behavior may be labeled teamwork, acting beyond the call of duty, selfless dedication, heroism and unshakable loyalty. Contextual performance not only helps the organization achieve its core purpose but also contributes significantly to other important outcomes such as positive work environment, low turnover, healthy recruitment and innovation. These outcomes are critical to growing and sustaining an organization.

Counterproductive Work Behavior

In contrast to contextual performance, "Counterproductive Work Behavior" or CWB in short (Spector et al., 2006; Spector & Fox, 2010) deliberately harms coworkers, superiors, customers or the organization. CWB behaviors include withdrawal (e.g., absenteeism, arriving late and leaving early), theft, abuse and sabotage. In military jargon, CWB includes behaviors such as AWOL, ragging, mutiny, fragging, committing atrocities against civilians and treason. While some of these behaviors may be related to task performance, most are carried out to directly hurt the targeted person(s) or the organization. The fragging of superiors (i.e., killing them using fragmentation grenades) in the U.S. Army (Bond, 1976) and the 1968 My Lai atrocities on civilians by U.S. soldiers (Peers, 1979; see Box 3-1) during the Vietnam War stand as extreme examples and a grim reminder of what soldiers are capable of when there is a confluence of precipitating conditions. A recent reminder of CWB is the 5th November 2009 U.S. Fort Hood shooting incident (McFadden, 2009) in which Major Hasan, an Army psychiatrist, shot and killed 12 soldiers and one civilian and injured as many as 30 others inside the camp. In the military workplace, another noticeable category of CWB is causing harm to oneself. This includes substance and alcohol abuse (and attempted suicide) which render the individual incompetent, unpredictable, out-of-control or unavailable, in short, unsuitable for task performance.

A Framework of Soldiers' Performance

By placing task performance on the y-axis and contextual performance-CWB on the x-axis, a framework of soldier performance is proposed (see Figure 3-1) with three distinguishable zones:

FIGURE 3-1. A Framework of Soldier Performance.

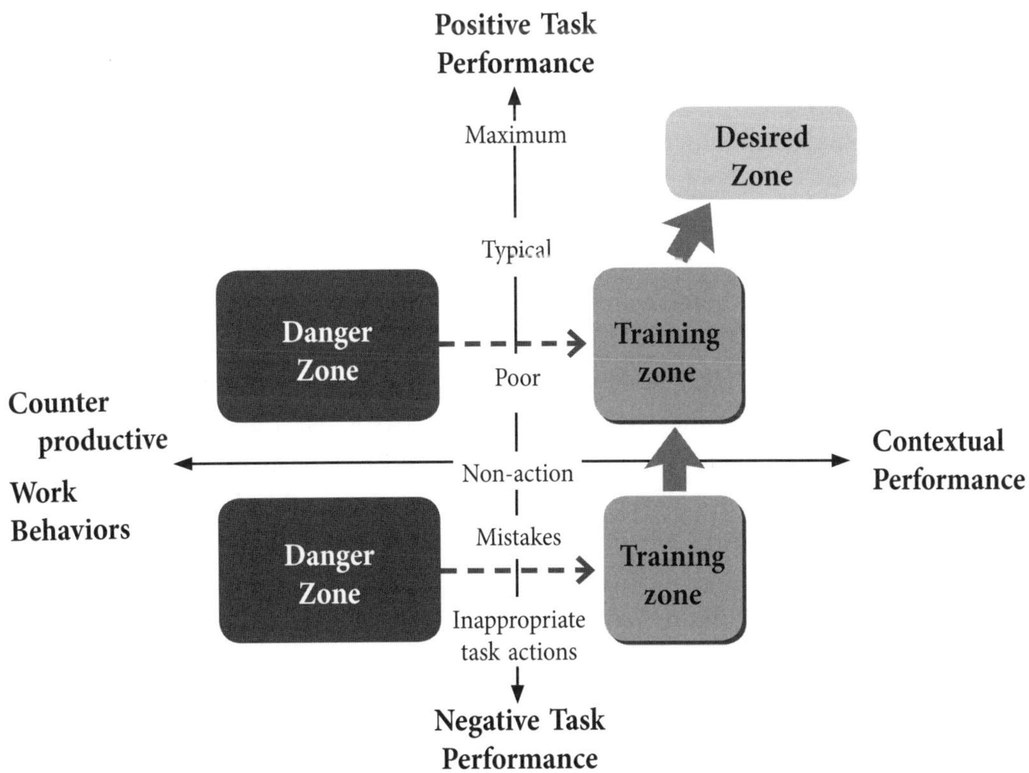

desired zone, training zone and danger zone. The desired zone is one in which soldiers put in their best effort and complete the task well, and while they are executing the task, they are encouraging one another and covering one another's lapses or gaps in performance. The training zone is one in which task performance ranges from negative to average, and where contextual performance could be further improved. The danger zone is where soldiers intentionally engage in CWB. In this zone of soldier performance, soldiers may perform according to expectations (i.e., show positive task performance) but their intentions and corresponding behaviors harm other individuals or the organization. These soldiers become liabilities as they pose a danger to others as well as the organisation. A soldier may be in any of the three zones depending on the situation and task assigned.

The framework allows soldier performances to be categorized and it serves to remind all commanders of the wide range of soldier performance behaviors that impact the mission, operation, unit and organization. The framework instructs commanders on what to monitor, what to work towards (i.e., moving soldiers from the training zone to the desired zone) and what to prevent or stop from happening (i.e., the slide from training zone to danger zone). To do so, one needs to have a good understanding of the factors that explain the four categories of soldier performance in the framework. The next section of this chapter will look at the psychology and patterns underlying performance-related behaviors.

THE PSYCHOLOGY OF HUMAN BEHAVIOR

Psychology has been commonly defined as the scientific study of human behavior and mental processes, which is usually taken to include observable actions as well as internal processes and responses of a person such as his/her thoughts, emotions and motivation. Among the fields of psychology, there are different perspectives of the psychology of human behavior. In the context of military operations, the social psychology and industrial-organizational psychology perspectives are commonly adopted in the study of soldier performance.

Based on the social psychological perspective, the behavior of an individual is the result of an interaction between him/her and the social situation, in particular the real or imagined presence and actions of other people (Aronson, Wilson & Akert, 1997). Social psychology provides a valid explanation for why individuals obey a leader's orders to carry out unethical behavior or conform to the norms in a team or behave differently when they are in a crowd. Based on the industrial-organizational psychology perspective, the behavior of individual is the result of the interaction between the work environment and the individual. Industrial-organisational psychology studies provide valid theories and explanations for questions like why different soldiers react differently to the same moral dilemma, why soldiers respond differently to the same stressors, and why soldiers are willing to give their best and ultimately sacrifice their lives in their jobs.

Behavior and Its Determinants

Behavior = Situation × Person

Behavior is essentially a function of the situation and the person and the generic equation is given above. Behavior is often defined as an action, and where specified, it may include the responses within the person and would involve thoughts/cognitions and emotions. Further, the equation focuses on conscious intentional behavior rather than subconscious learned behavior (e.g., experienced driver driving a car) or instinctive behavior (e.g., covering our ears when there is a loud noise).

The equation could be applied to the discussion on stress in the military as discussed in the previous chapter. In the model of stress, the "behavior" is represented by the "stress outcomes"; "situation" is represented by the "stressors" and the "moderators"; "person" is represented by the "appraisal" and "individual traits and prior experience moderator". The previous chapter also discussed one class of situational factors, i.e., challenges and stressors that can affect soldier performance. The next section discusses other situational factors related to soldier performance.

SITUATIONAL FACTORS

How situations affect behaviors is a key topic of study in social psychology. In this field, the dominant situational factors have to do with the people in the situation, e.g., the leader, peers, the team/unit, and strangers in a group/crowd. In the field of industrial-organizational psychology, some of the dominant situational factors include the mission/task demands, the environmental stressors and the unit/organizational climate and culture.

Studies in social psychology have enhanced the understanding of how soldiers behave and perform in social situations. In the following sections, key phenomena and findings, which were taken from psychology (Coon & Mitterer, 2010; Kalat, 2011) and social psychology textbooks (Aronson, Wilson & Akert, 1997; Baumeister & Bushman, 2011; DeLamater & Myers, 2011), and extracts of the analysis of the My Lai and Abu Ghraib incidents, will be highlighed to illustrate the power of situational factors (besides stressors) in influencing soldier behavior and performance.

The Power of Authoritative Figures to Elicit Compliance

In a well-known experiment (in terms of its interesting findings and the ethics of the study) conducted by social psychologist, Milgram, in 1963 (as cited in Aronson, Wilson & Akert, 1997; DeLamater & Mayers, 2011), Milgram succeeded in getting his subjects to inflict substantial "pain" on others by

BOX 3-1

The My Lai Massacre of Civilians during the Vietnam War in 1968

Vietnamese civilians indiscriminately killed by U.S. soldiers in My Lai in 1968 during the Vietnam War.
Source: Ronald L. Haeberle/Time & Life Images/Getty Images

- In March 1968 during the Vietnam War, Charlie Company of 1st Battalion, 20th Infantry Regiment, 11th Brigade, 23rd Infantry Division, was tasked with the mission to search and kill all North Vietnamese combatants and "suspects" in the My Lai village. When they entered the village, they did not find any North Vietnamese combatants. Soldiers from the 1st platoon, led by 2nd Lieutenant William Calley, began to attack the unarmed civilians.
- 2nd Lieutenant Calley ordered his soldiers to shoot at the civilians and he himself participated in a number of the killings. Some soldiers from the other platoons joined in the massacre. Although a number of soldiers did not participate in the atrocities, they did not protest nor report the incident to their superiors.

- It was estimated that over 300 civilians (mainly women, children and elderly people) were abused, tortured and killed with firearms, grenades and bayonets.
- During the court martial, 2nd Lieutenant Calley claimed he was following orders from his superior Captain Medina who was acquitted of all charges. In 1971, 2nd Lieutenant Calley was convicted of premeditated murder and sentenced to life imprisonment. However, he only served four-and-a-half months in a military prison. Most of the soldiers who were involved were enlisted men and had left the service by then and thus could not be prosecuted.
- An inquiry, led by Lieutenant General Peers, identified 13 factors that contributed to My Lai. Peers later combined some of them into 9 factors in his 1979 book. They are:
  - *Lack of proper training.* The soldiers were not trained in law and rules of engagement regarding non-combatants.
  - *Attitude toward Vietnamese.* The soldiers saw the Vietnamese as an inferior race and a lower form of human beings.
  - *Permissive attitude.* The division and brigade HQs condoned incidents of rough handling and mistreatment of prisoners, thus sending a wrong message to the soldiers.
  - *Psychological factors.* The soldiers were frustrated that their comrades had been killed and injured by mines and booby traps. They also had little direct contact with the enemy. There was also pressure from their task force commander to act aggressively and make contact with the enemy.
  - *Organizational problems.* Charlie Company was part of an ad hoc task force that had not worked together before.
  - *Nature of the enemy.* It was difficult to distinguish combatants from non-combatants during the Vietnam War.
  - *Plans and orders.* The orders were not clear and were distorted as they were passed from top to bottom, and soldiers were afraid to question them.
  - *Attitude of Vietnamese government officials.* The local officials believed that My Lai had combatants and sympathizers. Further, their attitude, which rubbed off on some U.S. soldiers, was that non-combatant sympathizers were expendable.
  - *Leadership.* Above all, a lack of leadership was the main cause as the leaders had created a unit climate that facilitated the massacre and failed to control the situation.

Source: From THE MY LAI INQUIRY by William R. Peers. Copyright © 1979 by W. W. Norton & Company, Inc. Used by permission of W. W. Norton & Company, Inc.

having an authority figure (a scientist dressed in white-coat) to instruct them to do so. The subjects thought that they were participating in a study on the effects of punishment in learning. The "learner", an accomplice of the experimenter, was strapped to a chair in an adjacent room. The subject was told to present problems to the "learner" to solve and to administer an electric shock each time a wrong answer was given. The subject was told to increase the voltage with each successive error, beginning with 15 volts to a high of 450 volts. Despite many subjects feeling distressed, 65 percent of the subjects proceeded to administer the final 450 volt level!

Three reasons were offered by social psychologists for the subjects' blind obedience to command: (1) people believe that their personal accountability for their actions are somehow reduced when there is a legitimate figure of authority physically present to issue the instructions and they assume the authority figure knows what he/she is doing, (2) the symbols and status of the authority figure and his/her institution (the white-coat scientist was from Yale University in the Milgram experiment) reinforce the validity of the experiment and the instructions from the authority figure and, (3) the gradual escalation of demands, beginning with seemingly innocuous orders, makes it difficult for subjects to find a definite point to declare what they were doing was wrong and to break away.

The experiment showed that individuals can be influenced to exercise destructive obedience, not unlike the obedience of soldiers who carried out the massacre of Jews during World War II and the massacre of civilians in the My Lai incident (see Box 3-1). Therefore, it is important for military training to prepare commanders and men to recognize and disobey unlawful orders.

Group (Informational) Influence on Individuals to Conform

A person is most likely to be influenced by the group (and not behave according to his/her own beliefs) if he/she believes that the group's behaviors or beliefs provide accurate information about reality or the correct way of behaving. This is especially so in ambiguous or crisis situations where there is no clear information, guidelines or norms for behaving appropriately. In such situations, if members of the group are unanimous about what they see, hear or believe to be right, the individual is most likely to accept the group's position and modify his/her own to align with the group. Where there is a lack of unanimity,

The behaviors of an individual in a group are often influenced by group consensus and norms.

Source: CyberPioneer, Ministry of Defence, Singapore

the individual is likely to be strongly influenced by those in the group that the individual perceives as experts or having experience on the matter of interest.

Group/Unit (Normative) Influence on Individuals to Conform

The group also exerts influence on the individual through group norms (shared expectations and standards of behavior) and beliefs. The pressure to conform is based on the phenomenon that people generally desire to fulfill others' expectations in order to be accepted and liked by others. The outward conformity may reflect sincere acceptance resulting in genuine change or it may simply reflect public acceptance. The extent to which a person conforms to the group norms, beliefs and demands depends on a number of factors such as the importance of the group to the individual, whether the group members are unanimous, the size of the group, the status of the group members relative to the individual, and the cohesion of the group.

The concept of group norms may be extended to the culture of the unit/organization where there are shared values, assumptions and understanding of what behaviors are appropriate and what are not. As with group norms, an individual will feel pressured to accept and conform to the unit/organization's culture especially if he/she desires to be a member. When an individual accepts and internalizes the shared values and assumptions of the unit/organization, they become part of his/her personal identity.

Both the My Lai and Abu Ghraib inquiries and analyses (see Boxes 3-1 and 3-2) indicated that unit/organizational climate and culture facilitated the soldiers' execution of their criminal acts. Climate (like mood and activities) and culture are, in turn, products of leadership and the responsibility of the commander. As such, commanders should be aware of the type of climate and culture they create in their units through their words, actions and inactions.

Role influence

Organizations and groups often have defined roles which are positions that come with an expected set of functions and behavior. While group norms specify how group members should behave, roles specify how individuals occupying those roles should behave within the organization or group. Roles can shape and alter individuals' behavior in powerful ways. This effect was clearly seen in the famous Stanford University Prison Experiment conducted by psychologist Zimbardo and colleagues in the 1970s (as cited in Aronson, Wilson & Akert, 1997).

A mock prison was built in the psychology department and paid students were randomly assigned to be either "guards" or "prisoners" and given the appropriate outfit and equipment. Very quickly, many of the "guards" became aggressive: they verbally abused and humiliated the "prisoners". Those who rebelled against the "guards'", treatment were punished and isolated from the others. The majority of "prisoners" became passive and withdrawn. A few became so stressed and depressed that they asked to be pulled out of the experiment. The experiment, which was to run for two weeks, had to be terminated after only six days. One conclusion was that the individuals' personal identities, associated self-controls and behavior became overshadowed by the roles which they adopted. In the military, there are many roles that influence soldier behaviors. The "leadership" or "command" roles are important and dominant roles that will influence how soldiers behave in their "follower" roles.

Crowd Influence: Deindividuation Effect

Another situational phenomenon from social psychology studies which may explain soldier behavior or performance is the phenomenon of "deindividuation effect". Deindividuation refers to the feeling of anonymity and loss of uniqueness experienced by individuals in a large crowd or a crowd of strangers. Researchers have found that deindividuation can lead to a relaxation of the normal constraints or social control on behavior. Deindividuation may explain why some individuals commit impulsive and deviant acts that they would otherwise not do on their own. Examples include mob behavior and hooliganism in soccer games, protests and other emotionally charged large gatherings. When an individual feels that he/she cannot be identified, he/she may feel less accountable for his/her actions which increases the willingness of that individual to engage in more extreme versions of the behavior. Another social psychology explanation is that the presence of a crowd lowers self-awareness and therefore self-control. The acts of others also provide justification for one to join in. Commanders could influence their soldiers' behaviors and performance by building and developing positive identities that can guard against the effect of deindividuation. Deindividuation will unlikely take place if soldiers feel that they are not strangers in a crowd but belong to a unit or group with an identity that represents certain values or virtues to be upheld.

Crowd (Bystander) Influence: Diffusion of Responsibility Effect

The "bystander effect" is another well known phenomenon in which the situation plays an important role in influencing behavior. On March 1964, "a woman named Kitty Genovese was attacked by a knife-wielding rapist outside her apartment in Queens, New York. News reports said her screams for help aroused 38 of her neighbors. Many watched from the windows while, for 35 minutes, she tried to escape. None called the police or sought help in any other manner." (Baumeister & Bushman, 2011, p. 278). The incident became national news and ignited a string of research. Although some of the neighbors reported that they thought it was a lover's quarrel, others cited reasons such as the impersonal environment of urban life and desensitization to such incidents in New York City.

BOX 3-2

The Abu Ghraib prisoners abuse in Operation Iraqi Freedom

Private First Class Lynndie England holding a leash attached to a detainee at the Abu Ghraib prison in Baghdad, Iraq in late 2003.

Source: Associated Press

- In April 2004, news of physical, psychological and sexual abuse of prisoners in Abu Ghraib prison in Iraq by U.S. soldiers was released by the media.
- Investigations confirmed that soldiers of the 320th Military Police Battalion and other intelligence agencies had carried out a range of criminal abuses and torture on the prisoners. The acts included forcing the prisoners to strip naked and perform all kinds of sexual and humiliating acts on one another, threatening them with a loaded pistol and rape, and having military dogs intimidate and frighten them. Many of these acts were photographed and video-taped.
- Seventeen soldiers and officers were removed from duty. Eleven were charged, convicted by court martial and sentenced to military prison. The commanding officer, Brigadier General Janis Karpinski, who denied knowledge of the abuses, was reprimanded and demoted to the rank of Colonel.
- Bartone (2008) analyzed the incident from a social-psychological perspective and proposed that a number of situational and person factors contributed to the behavior of the soldiers.

Situational factors:

- Ambiguity in the chain of command.
- Laissez-faire leadership.
- Lack of training.
- Poor discipline.
- Psychological stressors (ambiguity, isolation, powerlessness, boredom and danger).
- Ambiguous rules of engagement, standards of conduct, laws, regulations and orders.
- Pressure from higher command levels to deliver.

Person factors:

- Personality traits and hardiness.
- Psychological development or "maturity".

References: Bartone (2008), Carter (2004) and Hersh (2004).

Nevertheless, researchers have found that the phenomenon exists elsewhere and have since termed this the "bystander effect".

Latane and Darley conducted a series of studies in the late 1960s and early 1970s (as cited in Baumeister & Bushman, 2011) and found that before someone intervenes in an emergency, he/she has to notice the event, interpret it as an emergency, assume responsibility, believe he/she has the appropriate assistance to provide and weigh the cost/risk of providing the assistance. One key finding was that the greater the number of bystanders, the longer it takes for someone to step in to help. This is because no individual feels a strong sense that it is his/her personal responsibility to take action, a phenomenon known as "diffusion of responsibility" which has come to be closely associated with the bystander effect. Both the bystander apathy and diffusion of responsibility concepts can explain a lot of what went wrong in the My Lai and Abu Ghraib incidents.

Weak and Strong Situations

The earlier sections describe how situations might play a part in determining soldier behavior and performance. To further understand why some situations play a bigger or lesser role, Mischel (1968) introduced the concept of the strength of situations, i.e., the availability and clarity of cues in the environment regarding the expected behavior, in determining how much a situation influences the individual expression of his/her personality (and values, etc.). In "strong situations" where there is clarity on how one should behave (e.g., the way recruits walk and talk during basic military training), the relationship between the person factors and behavior is reduced. This means that person factors do not predict differences in individual behavior well, as individuals generally behave similarly because of the strong demands of the situation. In a "weak situation" where there is much ambiguity or less expectations about how one ought to behave, the opposite is true. For example, in informal social gatherings, individuals tend to behave according to who they are (person factors) and greater differences between individuals will be seen. The military example of a weak situational context is the "battlefield" where there is much ambiguity and uncertainty as opposed to the barracks (strong situational context) with regimentation governing how soldiers should behave. When one takes into consideration the effects of strong or weak situations, one might come to very different conclusions about the person than if one infers only from the behaviors. Strong or weak situations have implications for how one might behave. If one fails to take into consideration strong or weak situations, one is likely to commit the "fundamental attribution error".

Fundamental Attribution Error

The fundamental attribution error is the tendency to explain others' behavior entirely in terms of their disposition and traits, thus underestimating the power of social influence. It has been proposed that the phenomenon is due to the perceptual saliency of the persons, i.e., one is more able to physically observe and thus notice individuals as opposed to the often unknown or unobservable situational factors influencing their behavior. The explanation is strengthened by findings that demonstrate that when one is asked to explain one's own behavior, one often makes attributions to the situation and less about one's own traits or personality, i.e., one does not commit the same fundamental attribution error about one's own behavior.

PERSON VARIABLES

There are many person factors or variables studied by psychologists. These person variables explain differences between individuals in their responses to the same situation. This section briefly introduces some key person variables which influence work behavior and performance. Given that this is the domain of the soft sciences, there are overlaps among the variables and inconsistencies in the definitions among scholars. The following descriptions of variables are found in psychology textbooks and literature on personnel, social and work/industrial-organizational psychology. In particular, the following sources were referenced: Coon & Mitterer (2010), Franken (2002), Thornton (1992), Whetzel & Wheaton (1997), and Woods & West (2010).

Knowledge and Skills

The content of most military training programs focus on the acquisition of knowledge and skills by soldiers and leaders to perform relevant military tasks. "Knowledge" is essentially the organized information a person has. Knowledge can be further classified into different types according to different theories. Knowledge is one key determinant of most task performance, i.e., there is a positive relationship between knowledge and task achievement. All decision plans and execution of orders require some kind of knowledge. Knowledgeable commanders are seen as competent and they inspire confidence and trust in the soldiers they lead. As knowledge can be acquired through learning and experience, all commanders should aim to achieve a high level of knowledge in their professional domains.

"Skills' refer to sets of learned behaviors that have to be attained at a certain level of proficiency for a function or task. Examples include writing, swimming, driving and shooting. Like knowledge, skills are important determinants of task performance and can be acquired through learning and experience. As with knowledge, a skillful commander inspires confidence and trust in the soldiers he/she leads.

Abilities and Competencies

"Ability" has many definitions, ranging from narrow to broad. One definition of ability refers to an individual's innate capabilities to perform a function or task. Examples of such abilities include cognitive abilities such as intellectual ability (or intelligence), spatial ability and verbal ability. Ability refers to a general capability for a specific task (e.g., ability to run 2.4 km under 10 minutes) or a general function (e.g., ability to communicate). In the last couple of decades, human resource personnel have preferred to talk about competencies which refer to a broad and complex set of inter-related knowledge, skills and abilities for a particular function or role. For example, the competencies required of a commander may be leadership, management and communication.

"Competencies" may be elaborated as comprising certain skills and abilities, as seen in a number of military leadership models such as those of the U.S. Army and Singapore Armed Forces. Due to the imprecise definitions, what one model lists as a competency may appear as an ability or skill in another. As with knowledge and skill, competency plays an important role in influencing task performance and can be acquired through learning and experience. Like knowledge and skill, a competent commander inspires confidence and trust in the soldiers he/she leads.

Beliefs and Values

A belief is generally defined as something which one regards or accepts as true. Beliefs that are taken for granted and which one is not consciously aware of are known as "assumptions". Assumptions and strong beliefs have a powerful influence over behaviors. For example, the assumptions and beliefs a commander has about his/her soldiers' motivation and abilities could influence what leadership style or method of training he/she would adopt to manage them. Beliefs are formed and continuously reinforced or shaped by reading, listening, observation and experience. One also forms beliefs about oneself. One's beliefs become part of one's self-identity. Self-identities strongly influence how one behaves and one would usually strive to behave in a manner that is consistent with one's self-identity especially if it is strong.

A value is a type of belief about what is important in one's life. Values may be broadly divided into two types: values about what is important to attain (e.g., happiness, success, power and money) and values about how one should go about attaining desirable goals (e.g., being honest, responsible and compassionate). Moral values refer to what a society holds to be right or wrong with regard to human interaction and behavior. A person's value system refers to his/her internal hierarchy of values based on the strength of the values.

Like beliefs, values have a strong influence on an individual's behavior. The influence is exerted when one makes decisions about what to pursue in life or about how to respond to a situation or dilemma. In an uncertain and ambiguous "weak" situation, a soldier's value system will guide his/her behavior. For example, when a soldier is left with another injured fellow soldier on the battlefield with an advancing enemy, does he/she flee on his/her own or does he/she help the injured soldier but risk being slowed down and captured? Even in a "strong" situation where there is pressure to either conform to the leader's unlawful orders or a group's criminal suggestion, it is believed that a soldier who is imbued with the right values will do the right thing and not succumb to such pressures. Values therefore can have a strong influence on soldier performance, whether it is task or contextual, as well as on counterproductive work behavior. It is for this reason that the military and many organizations spend time and effort to build a strong culture and inculcate in newcomers the values of the organization.

Parents, guardians, friends, the school, employers, organizations, politicians and the media all potentially play a part in shaping an individual's personal beliefs and values. As people are embedded in a country or society, the culture of the country and society has an influence on the personal beliefs and values of the citizens and members. Once assumptions and values are formed, they become fairly stable and enduring. It often takes a strong experience and life-changing event for one to review and reorder one's value system.

Attitudes

An attitude is an enduring evaluation (positive or negative) of people, objects or ideas (collectively known as "target"). For example, people have attitudes toward their politicians, their armed forces

Values guide decisions and behaviors, especially in dilemmas and ambiguous situations.
Source: Corbis/ Click Photos Pte. Ltd.

and the idea of conscription. Some of the attitudes might be cognitively based, i.e., the evaluation is based on information and beliefs about the merits and demerits of the target. For example, "This politician is not good because he continued with his vacation despite the crisis back home." Other attitudes might be affectively based, i.e., the evaluation is based on emotions and values rather than an objective appraisal of the merits and demerits of the target. For example, "Conscription is great. I had a wonderful time when I was serving in the Army." Attitudes can change from negative to positive or vice versa as a result of direct experience or persuasion from others.

Attitudes have different strengths. A strong attitude is typically formed from a direct experience with the target. The stronger an attitude, the greater the influence on the perception, sense-making and behavior of a person toward the target. When a person has a prior attitude toward a target, it could affect how new information regarding the target would be perceived and interpreted. For example, if a commander has a negative attitude toward a soldier, the attitude could direct the commander's attention to the negative aspects of the soldier's behavior in a situation or lead him/her to interpret the soldiers' behavior in a negative way. Either way, it reinforces the prior attitude and results in a response that would be quite different from when he has a positive attitude toward the target. As the target could be a task, the enemy, superior, peers, subordinates or an organization, attitudes potentially have a strong influence on soldiers' task performance, contextual performance and counterproductive work behaviors. Adolf Hitler's prejudice against the Jews is a classic example of the destructive power of a negative racial attitude.

Personality

Personality refers to a distinctive pattern of behaviors (including thoughts and emotions) of a person. It is the totality of the attributes of a person that make him/her unique. One popular view of personality is that it consists of traits, i.e., consistent patterns of behaviors. Traits are regarded as underlying causes of many behavioral expressions. For example, an introverted person would rather stay at home and read a book than network with various individuals at a party. Many regard traits and personality as fairly innate, stable and enduring. As such, personality is often cited as the reason why individuals might behave consistently across different situations. This is the reason why some organizations believe in assessing people's personality in order to select the right applicants or screen out applicants whose personalities do not match the job requirements.

Notwithstanding the difficulty in defining a universal set of traits and, hence, the numerous personality theories and trait labels that exist in the marketplace, individuals do quite consistently differ from one another and they do exhibit fairly consistent behaviors across situations. For example, some commanders are naturally calm and collected while others have to work hard to control their anxiety. Some commanders are naturally warm and approachable while others are more aloof and emotionally detached. Under stressful and sleep deprivation conditions, self-controls and effortful concealment could be stripped away to reveal the core personality of the commanders and soldiers. Given the wide range of personality traits and tasks, there would indeed be some traits that could significantly influence a soldier's task performance, contextual performance and counterproductive work behavior. For example, a conscientious soldier would likely be reliable and dependable in his/her duties, a trusting soldier would likely come to the aid of his/her coworkers and organization, and a vindictive soldier would likely engage in counterproductive work behavior (e.g., sabotage) toward his/her accuser.

Motivation

Motivation is regarded as another underlying cause of behavior as it arouses and directs attention and energy for behavior. It is manifested in the choice of behaviors, its strength and persistence. For example, the motivation of a soldier will determine whether a soldier chooses to join or avoid an operation, how much strength and effort he/she puts in to train for the operation, and whether he/she throws in the towel when the conditions get tough.

Fundamentally, a soldier's performance hinges on two things — his/her task proficiency (i.e., the knowledge, skills and abilities/competencies that he/she has for the task) and his/her motivation. In simpler terms, is he/she able and willing to do the job? Much of the routine training in the military is devoted to developing task proficiency and when that is achieved, it is fairly stable and enduring with periodic reinforcing practices. On the other hand, motivation is a factor that is complex, highly context-dependent and pliable. As such, it has been referred to as the "X" factor or force-multiplier that can determine the outcome of a battle. There are many historical records of troops who, despite being inferior in numbers and equipment, have stood against overwhelming odds to hold off or defeat the enemy. Because of the importance of soldiers' motivation for military commanders and leaders, an entire chapter (see Chapter Four) has been devoted to the subject.

SUMMARY

It is important for commanders to view soldier performance beyond task performance. This is because soldiers can exhibit other behaviors (contextual performance and counterproductive work behaviors) that can directly or indirectly affect a mission or the unit's/organization's goals. The framework of soldier performance was developed to remind commanders of the range of soldier performance behaviors which they are accountable for and have to manage as leaders. The framework highlights what to monitor, what to work toward (moving soldiers from the training zone to the desired zone) and what to prevent or stop from happening (slipping from training zone to danger zone).

As performance is about behavior directed toward a task or an organizational goal, the psychology of human behavior provides an understanding of soldier performance. Behavior is basically a function of the situation and person. Situational factors such as an authoritative figure, the group's informational and normative influence, role influence and crowd influence can cause a person to behave in ways he/she would not normally do on his/her own. In this regard, the My Lai massacre of civilians during Vietnam War and the Abu Ghraib prisoners' abuse highlighted some important situational factors and lessons for commanders. Often, the power of the situation to influence behavior is underestimated and this is reflected in the "fundamental attribution error". Person factors or variables become more influential in a "weak" situation where there is much ambiguity and less expectations about how one ought to behave. Key person variables include knowledge, skills, abilities, competencies, beliefs, values, attitudes, personality and motivation. Each of them has the potential to influence one or more dimensions in the soldier performance framework. It is, therefore, imperative that all commanders are well-informed about the factors underlying soldier performance so that they can effectively manage most of these factors to enhance the performance of their soldiers to meet their units' mission or objectives.

Key Concepts

- Performance
- Task performance
- Contextual performance
- Counterproductive work behavior
- Behavior
- Psychology
- Group informational influence
- Group/unit normative influence
- Role influence
- Deindividuation effect
- Bystander effect
- Weak and strong situations
- Fundamental attribution error
- Knowledge
- Skills
- Abilities
- Competencies
- Beliefs
- Values
- Attitudes
- Personality
- Motivation

Questions

- How likely is a My Lai or Abu Ghraib type of incident to happen in your military?
- How well would your unit or any unit in your battalion measure up to the factors described in Boxes 3-1 and 3-2 when faced with the stressors of operations?
- How would you deal with the factors that you have assessed to be of concern?
- How well do you know your soldiers and yourself? Describe them and yourself based on the key person variables.

References

Aronson, E., Wilson, T.D., & Akert, R.M. (1997). *Social Psychology* (2nd ed.). New York: Longman.

Bartone, P.T. (2008). Lessons of Abu Ghraib: Understanding and preventing prisoner abuse in military operations. *Defense Horizons, 64,* 1–8.

Baumeister, R.F., & Bushman, B.J. (2011). *Social Psychology and Human Nature* (2nd ed.). Belmont, CA: Wadsworth.

Britt, T.W., Castro, C.A., & Adler, A.B. (Eds.). (2006). *Military life: The psychology of serving in peace and combat (Vol. 1): Military performance.* Westport, CT: Praeger Security International.

Bond, T.C. (1976). The why of fragging. *American Journal of Psychiatry, 133(11),* 1328–1331.

Borman, W.C., & Motowidlo, S.J. (1993). Expanding the criterion domain to include elements of contextual performance. In N. Schmitt & W.C. Borman (Eds.), *Personnel selection in organizations* (pp. 71–98). San Francisco: Jossey-Bass.

Borman, W.C., & Motowidlo, S.J. (1997). Task performance and contextual performance: The meaning for personnel selection research. *Human Performance, 10,* 99–109.

Campbell, J.P., McCloy, R.A., Oppler, S.H., & Sager, C. E. (1993). A theory of performance. In N. Schmitt & W.C. Borman (Eds.), *Personnel selection in organizations* (pp. 35–70). San Francisco: Jossey-Bass.

Carter, P. (2004, November). The road to Abu Ghraib: The biggest scandal of the Bush administration began at the top. *Washington Monthly.* Retrieved Jan 2, 2011, from ttp://www.washingtonmonthly.com/features/2004/0411.carter.html.

Coon, D., & Mitterer, J.O. (2010). *Introduction to psychology: Gateways to mind and behavior* (12th ed.). Belmont, CA: Wadsworth.

Dalal, R.S. (2005). A meta-analysis of the relationship between organizational citizenship behavior and counterproductive work behavior. *Journal of Applied Psychology, 90,* 1241–1255.

Deci, E.L., & Ryan, R.M. (1985). *Intrinsic motivation and self-determination in human behavior.* New York: Plenum.

Deci, E.L. & Ryan, R.M. (2000). The "what" and "why" of goal pursuits: human needs and the self-determination of behavior. *Psychological Inquiry, 11*, pp. 227–268.

DeLamater, J.D., & Myers, D.J. (2011). *Social Psychology* (7th ed.). Belmont, CA: Wadsworth.

Driskell, J.E., & Salas, E. (1996) (Eds.). *Stress and human performance*. Mahwah, NJ: Lawrence Erlbaum Associates.

Franken, R.E. (2007). *Human motivation* (6th ed.). Belmont, CA: Wadsworth.

Hancock, P.A., & Szalma, J.L. (Eds.). (2008). *Performance under stress*. Burlington, VT: Ashgate Publishing.

Hersh, S.M. (2004). *Chain of command: The road from 9/11 to Abu Ghraib*. New York: HarperCollins.

Kalat, J.W. (2011). *Introduction to psychology* (9th ed.). Belmont, CA: Wadsworth.

McFadden, R.D. (2009, November 5). Army doctor held in Ft. Hood rampage. *New York Times*. Retrieved Jan 2, 2011, from http://www.nytimes.com/2009/11/06/us/06forthood.html?pagewanted=1.

Mischel, W. (1968). *Personality and Assessment*. New York: Wiley.

Organ, D.W. (1997). Organizational citizenship behavior: It's construct clean-up time. *Human Performance, 10*, 85–97.

Peers. W.R. (1979). *The My Lai Inquiry*. New York: W.W. Norton and Co.

Rielly, R. (2009). The inclination for war crimes. *Military Review, 89*, pp. 17–23.

Sackett, P.R., Zedeck, S., & Fogli, L. (1988). Relations between measures of typical and maximum job-performance. *Journal of Applied Psychology, 73*, 482–486.

Spector, P.E., & Fox, S. (2010). Counterproductive work behavior and organizational citizenship behavior: Are they opposite forms of active behavior? *Applied Psychology: An International Review, 59(1)*, 21–39.

Spector, P.E., Fox, S., Penney, L.M., Bruursema, K., Goh, A., & Kessler, S. (2006). The dimensionality of counterproductivity: Are all counterproductive behaviors created equal? *Journal of Vocational Behavior, 68(3)*, 446–460.

Thornton, G.C. (1992). *Assessment centers in human resource management*. Reading, MA: Addison-Wesley.

Whetzel, D.L. & Wheaton, G.R. (1997) (Eds.). *Applied measurement methods in industrial psychology*. Palo Alto, CA: Davies-Black.

Woods, S. & West, M. (2010). *The psychology of work and organizations*. Belmont, CA: Wadsworth.

4

Understanding Soldier Motivation and Morale

Source: CyberPioneer, Ministry of Defence, Singapore

OVERVIEW

In the previous chapter, the basic theories and knowledge of human behavior were presented to explain the factors (situation and person) that influence soldiers' behavior and performance. Many situational and person factors were described and their impact on human behavior and soldiers' performance were illustrated. This chapter will focus on one of the "person" factors – motivation. Even though soldiers can become competent in their area of expertise through years of training, a great variability in their performance may still exist as a result of differences in their motivational levels. As such, soldiers' motivation or morale (a closely related concept) has been regarded as the "X" factor or force-multiplier in a combat situation and a key concern of military leaders, who seek to inspire and motivate soldiers to give their best and even sacrifice their lives for their nation and comrades if needed.

The first section defines motivation and presents a few well-known motivational theories. Many of these theories can be applied to soldiers in situations during peacetime, in training and even in operations. The second section focuses on combat motivation — what motivates soldiers to fight and give their best in combat, and examines the dynamics of motivation during the phases of an operation. The final section examines the concept of "morale", its similarities and differences with combat motivation, and how commanders can assess morale and use morale data or information for making decisions.

WHAT IS MOTIVATION AND TYPES OF MOTIVATION

Definition and Nature of Motivation

Motivation is often regarded as the underlying cause of behavior as it arouses and directs attention and energy for behavior (Franken, 2007; Petri & Govern, 2004). It is manifested in the choice or direction, strength and persistence of the behavior. For example, whether a soldier chooses (i.e., directs energy) to help or ignore a fellow soldier, how much effort the soldier puts in to help, and whether the soldier gives up when the situation gets difficult, reflect the motivation of the soldier. Motivation is dynamic and highly situation-dependent. The soldier's motivation may change after he/she has decided to help his/her fellow soldier and there may be many factors that can affect the change and level of motivation. Some of these factors will become clearer when the theories of motivation are presented later. In recent years, researchers have recognized the important role of emotions and are studying its relationship to motivation. Motivation is therefore complex and no one theory adequately explains this dynamic concept of motivation. Nevertheless, the theories still provide useful frameworks and models for commanders to use when they lead their soldiers.

Underlying Causes

Psychologists have identified two broad categories of underlying causes of human behavior — approach and avoidant. For the "approach" causes, people are motivated to do things because of something they need or desire. Examples include food, money and independence. For the "avoidant" causes, people are motivated to do things to prevent, avoid or escape something. Examples include failure, a bad smell and disapproval from loved ones.

THEORIES OF MOTIVATION

There are a number of theories of motivation and these are typically classified as "content" and "process" theories. The former focuses on the substantive causes of behavior which have been labeled as "needs" and "motives". Maslow's and Herzberg's theories fall under this category. The latter focuses on the process in which the behavior is energized and redirected. Reinforcement, expectancy and equity theories fall under this category. A brief description of each of the above theories is given below. They are based on the following sources: Franken (2007), Petri & Govern (2004) and Woods & West (2010).

Maslow's Hierarchy of Needs Theory: Our Needs Drive Our Behavior

Abraham Maslow (1908–1970) proposed that humans have various needs that drive behavior and that these needs exist on a hierarchy (see Figure 4-1) consisting of five stages, ranging from the "lowest", most primary biological and physiological needs to the "highest" self-actualization need. Needs at a lower level have to be satisfied before the next higher set of needs become relevant, i.e., people will be concerned about safety needs only after their biological and physiological needs (e.g., thirst, hunger, sleep and warmth) have been met. Likewise, safety needs have to be met before the need for belonging and love (which can be met by relationships both within and outside of the family) becomes a source of motivation. The next level of needs is esteem needs which is the need to achieve

FIGURE 4-1. Maslow's Hierarchy of Needs

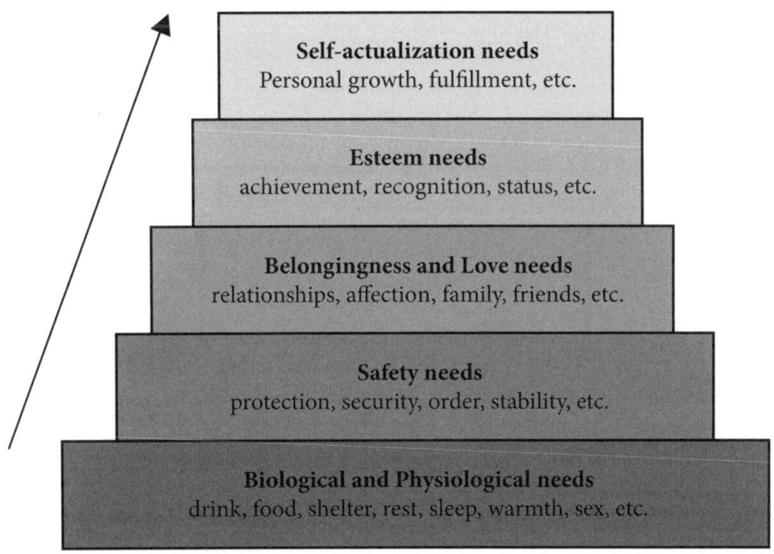

and see ourselves as competent as well as to be recognized and appreciated by others. Finally, at the top of the pyramid of needs is self-actualization, which is to pursue one's highest potential and to do the things one does best.

Maslow's theory which asserts that basic biological and physiological needs must be fulfilled before higher needs are pursued, is generally observed to be true. However, the levels of hierarchy and precise sequence have been criticized as rigid as higher order needs could precede lower order needs on some occasions and multiple needs from different levels could co-exist. Notwithstanding the criticisms, Maslow's theory serve to remind commanders not to underestimate the powerful basic needs of soldiers in the harsh conditions of the battlefield and to ensure the adequate supply of rations, to cater rest time and to resupply such basic needs. In the garrison, the model informs us that different incentives and approaches may be needed for different soldiers because they have different needs. Furthermore, incentives and approaches may lose their effectiveness after the soldiers' relevant needs have been met.

Herzberg Two-Factor Theory: Different Factors for Satisfaction and Dissatisfaction

Frederick Herzberg (1923–2000) proposed the "Motivation-hygiene theory", also known as the "Two-factor theory of job satisfaction". According to his theory, one's job dissatisfaction and satisfaction are influenced by two independent sets of factors: hygiene factors and motivators respectively (see Figure 4-2). Hygiene factors are generally aspects of the work environment and when they are

FIGURE 4-2. Herzberg's Two-Factor Theory

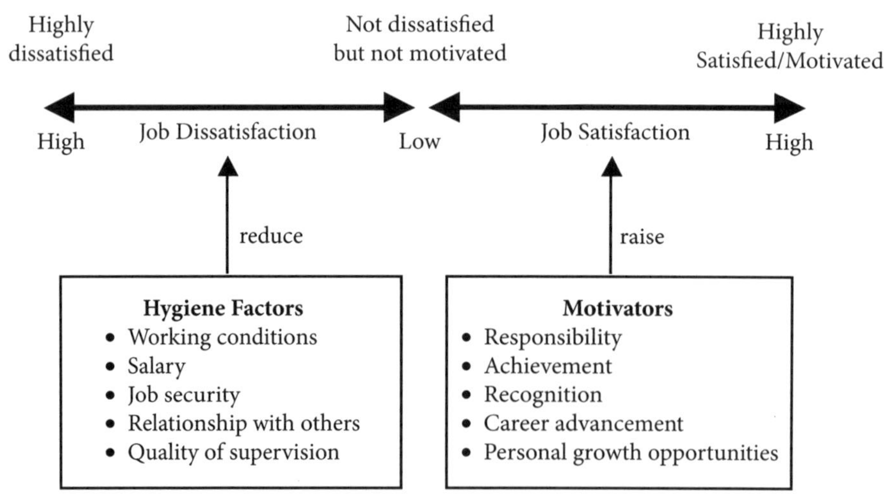

inadequate or perceived as poor, it results in employees feeling dissatisfied. When they are adequate, employees will not be dissatisfied but neither will they be particularly motivated at work. Motivators are mainly factors that relate to an employee's need for personal growth. When these factors are adequate, employees will be satisfied with their job and motivated to achieve high levels of performance.

There is some similarity between Herzberg's and Maslow's models. They both focus on the role of needs in motivation and mainly on the "approach" causes of motivation. These theories, however, do not explain in explicit terms, "avoidance" causes. In Herzberg's theory, only the higher levels of needs in Maslow's hierarchy (i.e., self-actualization and esteem needs) act as motivators. The remaining levels of Maslow's needs correspond to the hygiene factors and they can only cause dissatisfaction if they are not met. One of the criticisms about Herzberg's theory is that what might be hygiene factors for some people may be motivators for others depending on other contextual factors. For example, salary can be a motivator for people starting out in their careers whereas it could be a hygiene factor for someone in a highly paid senior position. Notwithstanding the criticisms, the Herzberg model is easy for commanders to understand and for commanders to discern hygiene factors from motivators (Some scholars refer to hygiene factors as extrinsic factors of motivation.) Hygiene or extrinsic factors are necessary for maintaining soldier morale and avoiding dissatisfaction while motivators or intrinsic factors need to addressed to be truly motivate soldiers to be satisfied and give their best.

Reinforcement Theory: Motivation via Carrot and Stick

B.F. Skinner (1904–1990) proposed the theory that behavior can be reinforced by external rewards. (External rewards here is used as a neutral term to refer to something given that can influence behavior). Positive rewards (i.e., incentives) increase the probability of a behavior and negative rewards (i.e., punishment) decrease its probability. Skinner showed that partial reinforcement (i.e., getting rewarded only some of the time rather than each time a desired behavior is produced, e.g., in gambling) can result in persistent behavior. Subsequently, researchers developed the theory further and explained that expectations of rewards following a behavior that is learnt from observing others can serve as reinforcement as well. Rewards need not always be from an external source, i.e., they can be internally generated as well. For example, "Achieving a task will prove that I am capable". Commanders are likely to be familiar with the concepts of this theory as positive rewards such as smiles, praises, rest and canteen breaks, and negative rewards such as reprimands, push-ups, extra training and confinements are often used during training to reinforce good performance behavior and punish poor performance behavior respectively. This motivational approach is based on the premise that soldiers' performances are largely a function of their motivation, that they have full control over their behavior, and that the incentives and punishments matter to them. The next theory will highlight some of the deficiencies in these assumptions.

Expectancy Theory: Motivation via Beliefs and Expectations

Expectancy theory was first formulated by E.C. Tolman (1886–1959) and subsequently expanded upon by V. Vroom in 1964. The theory (see Figure 4-3) assumes that behavior results from choosing a course of action among alternatives, with the aim of achieving valued outcomes, and to maximize pleasure and minimize pain. It has the following important components: valence (V), instrumentality (I) and expectancy (E). Therefore, the theory has been coined the VIE theory. An expectancy is a belief about the likely outcome of a particular behavior. The effort-performance (E-P) expectancy is the individual's expectation that his efforts will lead to the required performance level and may be quantified as a mathematical probability. Instrumentality is the individual's expectation about the extent to which the successful performance will lead to a particular outcome, and may also be quantified as the performance-outcome (P-O) probability. Valence is the perceived value or importance of the performance and outcome. Valence can range from +1 (what the individual strongly desires) to -1 (what the individual strongly wants to avoid).

In the theory, motivation is a function of the cross-product of Expectancy (E-P probability), Instrumentality (P-O probability) and Valence(s). If any of them is zero, there will be no motivation. For example, a soldier will not be motivated if he does not expect his effort to lead to the required performance level even if he believes that achieving the performance level will lead to a highly desired incentive. The theory is therefore more complex than reinforcement theory which simply informs commanders to administer incentives and punishments to motivate soldiers. Commanders have to work with soldiers' expectancies, instrumentality and valences. In order to improve the effort-performance expectancy, commanders should focus on training to improve the soldiers' abilities and their belief that their effort will indeed lead to better performance. Performance goals should be challenging but achievable. In order to enhance the performance-outcome expectancy, commanders

FIGURE 4-3. Vroom's Expectancy Theory

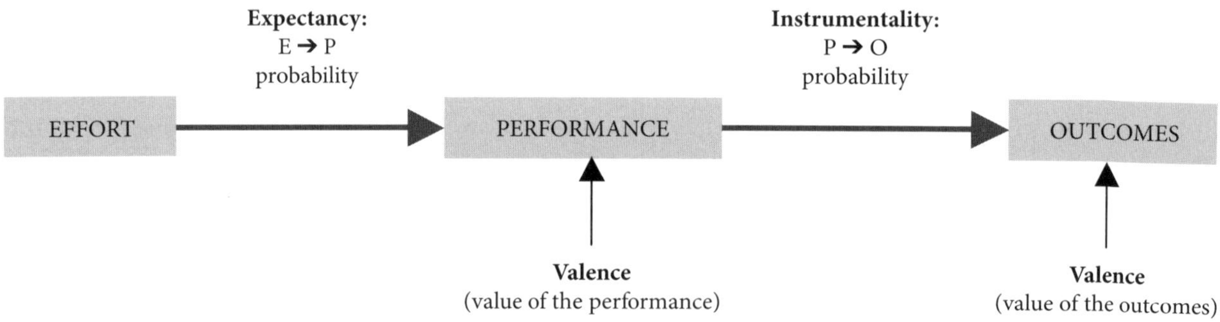

Motivation = Expectancy × Instrumentality × Valence (or VIE which is the popular acronym)

should ensure performance standards are clear and deliver on promised rewards as soon as possible after the standards have been achieved. Commanders also need to ensure that the rewards provided are desired (i.e., valued) by the soldiers.

The strength of the theory is that it takes into consideration the thought processes, values and beliefs of soldiers, thus providing an alternative explanation to the needs theories of motivation. Further, factors such as self-confidence, trust in leaders and ideology of war, which are found to be important in studies of combat motivation (to be discussed in this chapter later), may be partially explained by the theory in terms of expectancies and valence. The drawback of the theory is that it assumes that individuals make rational calculative choices all the time. The theory leaves no room for how emotions may affect the expectancies and valence, and disrupt the mental process.

Equity Theory: Motivation via Fairness Perceptions

It is natural for humans to compare. Equity theories, the first formulated by John Adams in 1965, propose how employees' perceptions and comparison of fair and unfair distribution of rewards, rather than the actual amount received, affect their motivation. Adam's theory asserts that employees are motivated to maintain equity between the inputs they bring to the job and the outcome they receive from it against the perceived inputs and outcomes of others in comparable jobs. Inputs are the employee's contributions in exchange for the outcomes. Examples are effort, time, skill, trust, loyalty and personal sacrifice. Outcomes may be positive or negative consequences and they may be tangible or intangible. Examples are salary, promotion, recognition, praises and reputation.

If an employee perceives his/her own ratio of outcomes over inputs to be less than the ratio of comparable others, he/she will perceive that there is inequity and injustice and that he/she has been

FIGURE 4-4 Equity Theory

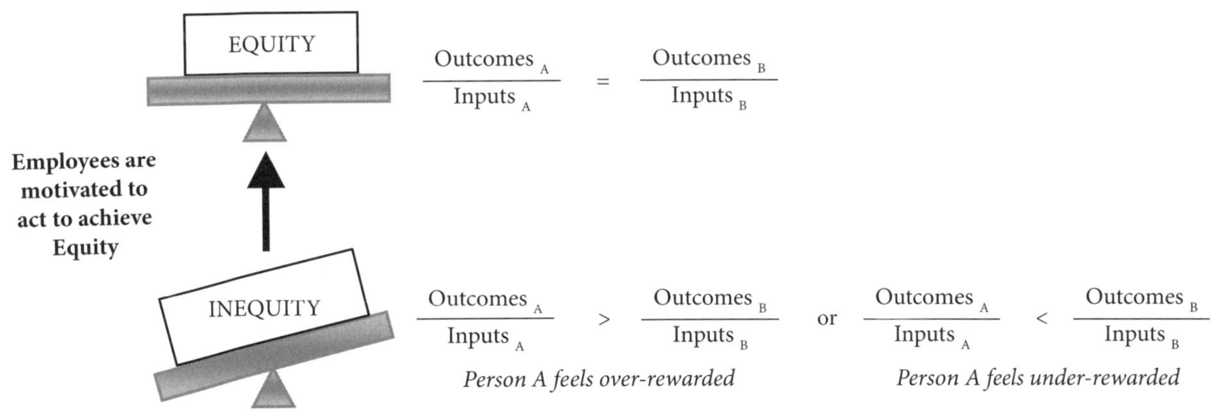

under-rewarded and disadvantaged. He/she is likely to feel angry or humiliated and such feelings may motivate him/her to reduce inputs, talk to the superior about the injustice or change his/her attitude toward the superior, peers and workplace. If the ratio is higher than others, there will be an inequity but this time, the inequity is in his/her favor (a case of being over-rewarded). He/she may feel guilty about getting more than his/her peers for what he/she had put in. He/she may then increase inputs or reduce his/her perception of the value of the outcomes.

Equity theory informs commanders about a number of important issues with regard to motivating soldiers. First, there are many different inputs and outcomes that soldiers may consider and not just what is given to them by the commander. A tangible outcome that is considered as an incentive by the commander might be less important to a soldier than to an intangible outcome. Second, it is not enough to just consider the magnitude of the absolute outcomes even if they are substantial, as soldiers will inevitably compare their outcomes with what others are getting. Third, commanders need to be aware of and actively manage their soldiers' perceptions of the inputs and outcomes and their comparisons through effective communication. Comparisons by soldiers are likely to happen, for example, when soldiers feel that their unit has been rostered for extra chores or dangerous duties and are not fairly rewarded. If commanders do not find out the relative importance of motivators, they may end up providing the ineffective motivators. At worst, a collective sense of injustice and accompanying anger can lead to serious motivational issues.

Self-Determination Theory: Moving from Extrinsic to Intrinsic Motivation

Theories of reinforcement, expectancy and equity typically deal with extrinsic motivation. Extrinsic motivation is based on external incentives and punishments that the behavior may bring. On the other hand, intrinsic motivation for a behavior is based not on any externally imposed source of influence or goals but rather on the value or pleasure associated with the behavior. While both types of motivations may be necessary for the same behavior, it is intrinsic motivation that is preferred. Intrinsic motivation initiates, self-regulates and sustains behavior. This class of motivational theory is known as self-determination theory. Research by Deci & Ryan (1985) suggests that all humans strive to satisfy the three innate and psychological needs of relatedness, competence and autonomy. This research supports to some extent Maslow's theory in terms of belongingness, esteem and self-actualization needs. In addition, Deci & Ryan's research expand on (a) the development of intrinsic motivation, (b) effects of stifling psychological needs (which lead to decreased motivation and learned helplessness, (c) how intrinsic motivation can be increased by giving more autonomy and positive feedback, or reduced by imposing tangible rewards, and (d) how extrinsic motivation can be transformed into intrinsic motivation. For the transformation to take place, the theory and data suggest that people need to feel that they are members of the group, identify with and internalize the task goal which means they need to see value in achieving an external goal and own it, believe they have the competency to perform and be given the autonomy to attain the goal.

Self-determination theory reminds commanders that soldiers are intrinsically capable of being self-motivated and self-regulated when they are well-trained and well-led. Intrinsic motivation is achieved not by giving more tangible incentives or threats of punishment but by building cohesion, esprit de corps and self-confidence as well as giving them a sense of purpose, which is the essence of good leadership and the subject of the next section on combat motivation.

COMBAT MOTIVATION — WHY SOLDIERS FIGHT?

Long history

Commanders have long realized that their soldiers' combat motivation or morale is the X-factor or force-multiplier on the battlefield for a successful outcome. "Not numbers or strength bring victory in war, but whichever army goes into battle stronger in soul, their enemies generally cannot withstand them" was a statement made by Xenophon, a Greek soldier, as early as 400 BCE (Manning, 1991, p. 453–454). Similarly, around the same time in the East, military strategist Sun Tzu in the *Art of War* (Griffith, 1963) also acknowledged the importance of soldier motivation in his emphasis on the moral cause for war, the quality of the soldiers and winning by overcoming the will of the enemy. Throughout the history of warfare, commanders, historians and researchers have spoken of combat motivation and given it various names such as "soul", "spirit", "moral strength" and "will-to-fight". A closely associated and perhaps more commonly used term by commanders is "morale". Is combat motivation and morale the same thing? This question will be examined in the next section.

What is Combat Motivation? What are the Underlying Factors?

Combat motivation is about the motivation of soldiers to engage in combat. As combat is about killing others or being killed; it involves both aggression and survival. Hence, combat motivation is defined as the reason and drive to inflict harm on others, i.e., to engage in a fight (choice), to fight hard (intensity) and to continue to fight (persistence) in the face of possible death. In the words of Anthony Kellet (1984), author of the classic and authorative book on combat motivation, motivation "comprises the influences that bear on a soldier's choice of, degree of commitment to, and persistence in effecting, a certain course of action" (p. 6).

At the psychological level, combat motivation can be viewed as a contest between fear, (which is to avoid or escape the fight), and the will to stay in the fight. As highlighted in Chapter 2, fear is a normal emotion in soldiers as they face threats of being killed, maimed or captured in an operation.

Thus, why soldiers fight and what keeps them going in the face of such threats has been the subject of study by researchers and historians. Kellet (1984) discussed a list of 13 major factors that influence combat motivation. The following paragraphs describe and highlight some of the key underlying factors and ideas from other scholarly works such as Britt & Dickinson (2006), Catignani (2004), Gal (1986a), Manning (1991), Rodrigues-Goulart (2006), Stouffler et al. (1949), Van der Dennen (2005) and Wong, Kolditz, Millen & Potter (2003). Given the dynamic nature of combat motivation and the challenges of conducting field research in this domain, much of the understanding of combat motivation comes from interviews, observations, inferences and conceptualizations rather than rigorous empirical studies.

Basic Biological and Physiological Needs

Maslow's hierarchy of needs theory informs us that all humans have basic biological and physiological needs such as food, drink, rest and sleep. When these basic needs for survival are not met, they can quickly drain the combat motivation of the soldier. When supplies are short, soldiers will not be vigilant, their minds will be distracted by hunger, thirst and discomfort and they will have less energy to last a fight. Herzberg's two-factor theory suggests that these needs correspond to hygiene factors, i.e., they are important to bring the soldiers out of misery and not destroy their combat motivation, but abundant provisions and supplies will not necessarily raise their combat motivation further. Soldiers will need motivators to increase or sustain their combat motivation for performance and mission success. Five of these motivators which are commonly cited in the military psychology literature are elaborated in the following paragraphs.

Cohesion

After World War II, Stouffler and his team of researchers (1949) interviewed war veterans about their attitudes and asked the following question: "Generally, in your combat experience, what was most important to you in making you want to keep going and do as well as you did?" Besides the most common answer which was that they wanted to get the war over and done with so that they could go home, they found that primary group cohesion, i.e., the strong emotional ties and bonds among the soldiers within their immediate team or unit, was the next most common answer and the main reason cited for combat motivation. This finding was also noted by others such as Marshall (1947) who wrote in his book *Men against fire* the following: "I hold it to be of the simplest truths of war that the thing which enables an infantry soldier to keep going with his weapons is the near presence or the presumed presence of a comrade."

In 2003, Wong et al. published a study on why U.S. soldiers fight in Operation Iraqi Freedom. Their aim was to examine whether the age-old belief and finding that team or unit cohesion was indeed

Cohesion — it provides soldiers the motivation to fight and face adversities.
Source: Scott Olson/Getty Images News/Getty Images

the most important factor that motivated soldiers to fight. The researchers asked the soldiers the same question Stouffer et al. used in their 1949 study and they similarly found cohesion to be the key psychological reason. Wong et al. proposed that team or unit cohesion serves two functions: firstly, it made soldiers responsible for the team and unit success and for protecting one another; secondly, it reassured them that they were not alone and that their team-mates were watching out for them. Others have added that soldiers fight because they did not want to let their comrades down — there was a sense of duty and loyalty to those whom they bonded with. Cohesion, besides being the reason why soldiers will fight and continue to fight in battles, was also found to be related to various performance and other positive outcomes as researchers have found (Oliver, Harman, Hoover, Hayes, & Pandhi, 1999; Siebold, 2006; Stewart, 1989). Some of these outcomes include a lower rate of absence-without-official-leave (AWOL) and desertion, mitigation against combat stress reaction and mental breakdowns on the battlefield and the facilitation of heroic acts by the soldiers.

Leadership

Leadership is another important determinant of soldiers' combat motivation. The influence of leadership on combat motivation is both direct and indirect. On the battlefield, leaders provide the direction and catalyst for action. In the fog of war, leaders are essential in leading confused and frightened soldiers. When leaders lead by example and take the initiative to charge or move when enemies are firing, their soldiers will follow. Sometimes this influence is instinctual. At other times, it is because soldiers have come to trust the leadership and decisions of their commanders though training together. When soldiers do not trust that their leader is competent or cares for them, they would not be confident in his decisions and be motivated to obey especially when their lives may be at risk. In the Vietnam War, there were reported cases of leadership failures in which soldiers turned on their officers and killed them with fragmentation grenades, an act known as "fragging" (e.g. Bond, 1976). To achieve success in motivating soldiers in training and combat, it is essential for commanders to inspire confidence and trust in their soldiers by being competent, caring and exercising leadership by example whenever appropriate.

Leadership also affects soldiers' combat motivation indirectly through other means. For example, good commanders exercise good leadership by building team and unit cohesion, unit esprit de corps and soldiers' self-confidence. Good leaders also shape their soldiers' beliefs, values and ideology with regard to the war or operation. When commanders exercise good leadership, they are likely to build well-trained, professional and motivated soldiers as well as effective units for combat.

Self-Confidence

Self-confidence is another key ingredient of combat motivation. Self-confident soldiers are more mentally prepared to engage the adversary and keep fighting because they believe in their own ability to perform and achieve mission success. In the language of expectancy theory, they have

Self-confidence enhances soldiers' motivation to fight.
Source: CyberPioneer, Ministry of Defence, Singapore

high E-P and P-O expectancies. They are not overwhelmed by anxiety and fear in carrying out their duties and tasks. A number of factors, such as training, leadership, weapons and equipment, and the image and actions of the enemy can affect soldiers' confidence. Good training increases the skills of the soldiers, teamwork and coordination with other units, thus strengthening their belief that they will defeat their adversary. Soldiers' confidence also depends on their confidence in their commanders (in their leadership), their weapons and equipment, and their military force's ability as a fighting system. Their confidence is also influenced by their perceptions of the enemy's combat power, weapon systems, technology, equipment, leaders and soldiers' proficiency and fighting spirit relative to their own.

Soldiers' self-confidence could be quite malleable and needs to be maintained. Adversaries may use psychological operations (including deception), playing up a military force's weaknesses and fears while they highlight their own strengths. Commanders need to be aware of these efforts and counter them with communication and actions to maintain their soldiers' confidence. It is inevitable that psychological operations and one's own forces' communication will influence soldiers' image of and attitude toward the enemy. While certain attitudes toward the adversary may be useful in securing soldiers' confidence, commanders need to guard against negative unintended consequences like in the case of the My Lai incident during the Vietnam War where over 300 civilians were massacred by U.S. soldiers (see Box 3-1 in Chapter 3). One of the contributing factors found in the U.S. Army inquiry was the negative attitude of the soldiers toward the Vietnamese people whom they saw as an inferior race.

Ideology

While one may think that ideology — the set of beliefs governing the soldiers' thoughts about the reasons for war and their involvement in it — will be a key factor for combat motivation, this presumption has not been a consistent finding in research studies. Stouffler et al. (1949) did not find ideology and patriotism to be major factors in the combat motivation of the soldiers they studied from World War Two. On the other hand, Wong et al. (2003) found many of the U.S. soldiers in Operation Iraqi Freedom to be motivated by patriotic ideals. More specifically, they commonly reported themes such as liberating the people and bringing freedom to the country. Wong and his team explained that the difference could be due to the U.S. soldiers being an all volunteer force and more educated and politically savvy. Another reason could be that the authors interviewed the soldiers while they were still actively deployed in the theater of operation — those who made it to Iraq and stayed on would not be conscientious objectors (i.e., persons who refused to serve as combatants or bear arms due to religious or moral beliefs) or have strong objections to the reasons for the operation.

Soldiers need an ideological cause or legitimate reason to hold on to when assaulted by scenes of death, destruction and cruelty in war. However, when soldiers are engaged in a firefight, their attention

Beliefs about the purpose of the war/operation affect a soldier's motivation to participate in it.
Source: Wathiq Khuzaie (left) and Jamal Aruri (right)/Getty Images News/Getty Images

will not be on such ideological thoughts but on their tasks at hand, their commanders' and comrades' actions and their duties to their comrades and unit. Therefore, the logical conclusion is that ideology and beliefs regarding the cause of war/operation are important to bring them to and keep them in the theater of operation, but are not sufficient to explain their combat motivation and performance during a firefight.

Esprit de Corps

While cohesion refers to the bonds in the primary group such as the team or unit, esprit de corps refers to "pride in and devotion to the reputation of a formal organization beyond the primary group" (Manning, 1991, p. 458). For the Army, "formal organization" or unit beyond the primary group might be the battalion, regiment, brigade or formation. For the Navy, it might be the ship or squadron. For the Air Force, it might be the squadron or base. When soldiers identify themselves with their unit beyond their primary group, their self-esteem is linked to the reputation of the unit or organization, and an observer would say that the forces have esprit de corps. When there is esprit de corps, soldiers take pride in the success of their unit or organization and share its pain and disappointment in failure. They contribute to and speak well of their unit or organization (much akin to the concept of contextual performance discussed in Chapter 3). Esprit de corps, therefore, provides additional motivation for soldiers to join the battle or operation, fight hard and stay on for the glory and reputation of their forces.

MORALE

Commonly Used but Lacks a Common Definition

Compared to combat motivation, morale is a term more commonly used by military leaders but less commonly agreed upon in its meaning. Combat motivation is easier to define because it is about motivation, which in itself has been well-studied and defined by psychologists in the last few decades, to do combat. Combat itself is also a fairly clear and narrowly-bounded domain. However, it is more common to find military leaders and the media talking about the level of "morale", rather than "combat motivation", of their troops and units in garrison, during training, and before, during and after an operation. As one can immediately see, morale is used more broadly than combat motivation to describe the troops or units in different environments.

The definitions and concept of morale were comprehensively reviewed by Baynes (1967), Manning (1991) and recently revisited by Britt & Dickinson (2006). Morale has been used by military leaders and researchers to refer to many different things about the soldiers. For example, past military generals and commanders have referred to it as the "soul", "spirit", "will" and "mental strength" of the soldiers. The 1983 U.S. Army field manual on leadership defined morale as "the mental, emotional, and spiritual state of the individual. It is how he feels — happy, hopeful, confident, appreciated,

High morale — the ingredient for high performance and psychological well-being of soldiers.

Source: CyberPioneer, Ministry of Defence, Singapore

worthless, sad, unrecognized, or depressed" (FM22-100, p. 228; Manning, 1991). In the latest edition of the field manual published in 2006 (FM6-22) (U.S. Army, 2006), there was no formal definition of morale but it was referred to as a "psychological force", "emotional element", "how people feel about themselves, their team, and their leaders", and "the essential human element — holds the team together and keeps it going in the face of the terrifying and dispiriting things that occur in war" (p. 7–8). The Australian Defence Force's (2007) doctrine on leadership acknowledged the difficulty of defining morale and avoided defining it by providing examples of definition. Similarly, the Canadian Forces' (2005) leadership doctrine did not directly define morale. On the other hand, NATO Research and Technology Organization, in its *A leader's guide to psychological support across deployment*, defined morale as "a service member's level of motivation and enthusiasm for accomplishing mission objectives" (NATO RTO, 2008, p. G-32).

Researchers also differ in their definition of morale. Motowidlo & Borman (1977) defined morale as "a psychological state shared by members of a group that consists of general feelings of satisfaction with conditions that have impact on the group and strong motivation to accomplish group objectives despite obstacles or adversity" (p. 177). Manning (1991) defined morale as "the enthusiasm and persistence with which a member of a group engages in the prescribed activities of that group" (p. 455). Britt & Dickinson (2006) defined morale as "a service member's level of motivation and enthusiasm for accomplishing mission objectives" (p. 162) which incidentally is the same definition as the one found in NATO RTO's (2008) guide for leaders. It was reported that most U.S. soldiers responded with the attributes of motivation and drive when they thought about morale. According to Brit & Dickinson (2006), the term morale could be used to describe both the individual and group (or unit) while other researchers (e.g., Catignani, 2004; Gal, 1986b) have viewed morale as a meaningful state of the group whereas combat motivation as a state of the individual.

In summary, definitions of morale have ranged from fairly broad and general (i.e., psychological and/or emotional states) to the narrow and specific (i.e., no difference from combat motivation) to somewhere in between (i.e., a psychological state/attitude which includes motivation). Fortunately, there appears to be some consensus among researchers and military leaders that for the construct to be useful, it must be more than a state of happiness, satisfaction, or feeling recognized or appreciated. Most researchers and military leaders will agree that morale should be evaluated with reference to the attainment of mission objectives or team and organizational goals. The shift in the U.S. Army leadership doctrine on the definition of morale reflects the consensus among the researchers on the topic. These two criteria are not trivial as they help to reduce the subjectiveness and enhance the usefulness in the monitoring, measurement and reporting of morale. For example, when the criteria are applied, simplistic statements such as "Don't worry, the morale of the troops will be high after I give them a canteen break and a day's rest" will not be valid. Such criteria about morale will also ensure that it is not seen as a transient mood state that does not inform commanders about soldier motivation prior to challenging missions.

Importance and Outcomes of Morale

The importance of morale is evident from the review of a sample of Armed Forces leadership doctrines/guides. For example, the word "morale" was mentioned 34 times in the main text of the U.S. Army manual FM6-22 on "Army Leadership" (2006). More importantly, the doctrine emphasizes the duty and responsibility of the leader to build and sustain the morale of his/her unit. Similar emphasis can be seen in the leadership doctrines and guides of NATO, Canadian Forces and Australian Defence Force. The NATO RTO (2008) guide for leaders went even further by devoting two-thirds of a chapter on how, what and when to measure morale, and what to do with the results subsequently (see Box 4-1).

Despite the lack of a common definition, there is a well-established assumption that high morale will lead to high performance and less psychological problems in the troops/units. The assumption is founded on the experiences and observations of past and present military leaders as well as some scientific studies. In spite of the inherent difficulties of conducting research in this area, researchers have generally found support for morale (as defined by the researchers) to be positively correlated with soldiers' or units' performance, psychological health and other positive outcomes (see Belenky, Noy & Solomon, 1987; Britt & Dickinson, 2006; Mannings, 1991; Motowidlo & Borman, 1977, 1978). For example, Motowidlo & Borman (1977) found that U.S. soldiers' self-report of platoon morale was significantly correlated with observers' ratings of the platoon's performance and effort on the job, and observers' ratings of morale of the platoons were significantly related to the soldiers' self-reported intent to re-enlist.

Factors Influencing Morale

Given the conceptual similarity between morale and combat motivation in many of the aforementioned definitions, it is not surprising to find commanders and researchers proposing the same factors for morale and combat motivation. In Britt & Dickinson's (2006) conceptual model of morale, there are four group factors with sub-group factors within each. They are: (1) mission-relevant factor with sub-factors such as clear purpose and achievable objectives, (2) leadership factor with sub-factors such as ability to clarify objectives and instilling high self and collective efficacy, (3) unit factor with sub-factors such as cohesion and collective efficacy, and (4) individual factor with sub-factors such as dispositional optimism, hardiness and self-efficacy.

There are some empirical studies to support the underlying factors proposed by Britt & Dickinson (2006). For example, using a survey questionnaire that is widely administered in the Israeli Defense Force (IDF) to measure unit combat readiness and morale, Gal & Manning (1987) examined the soldiers' report of personal and unit morale with several factors across two U.S. Army armored cavalry squadrons and a similar unit from IDF. They found that in the IDF unit, personal morale was most related to unit cohesion, self-confidence and relationship with their officers. Although there were

BOX 4-1

How to assess Morale and use the findings?

(based on NATO RTO (2008) guide for leaders)

How

- Leaders can assess the morale of their troops either informally by talking and listening to them in casual settings, or formally using focus groups discussions and surveys.
- Formal assessments have the advantage of allowing leaders to track changes systematically over time.
- Focus groups discussions should be facilitated by experienced facilitators who are not part of the chain-of-command with structured questions prepared in advance to emphasize certain issues.
- Successful morale surveys are jointly developed and conducted by leaders and military psychological support professionals who are trained in survey methodology. Surveys are more objective (as they are anonymous and all soldiers can participate with equal contribution) than focus groups but they often provide less information about the cause of a morale problem.

What

- Leaders might want to consider the following areas for focus group or survey:
  - Climate
  - Cohesion
  - Leadership
  - Efficacy
  - Stressors
  - Deployment events
  - Psychological health

When

- Leaders should consider a morale survey near the end of pre-deployment preparation, during-deployment (several times, if not, early in mid-phase) and post-deployment.

What to do with the survey findings

- The findings are meant to help leaders manage their units more effectively and should not be used to assess the leaders' performance.
- Military psychological support professionals can help in the objective interpretation of the findings and development of general recommendations.
- Leaders have an obligation to share the main findings with their unit members and resolve any morale issues. Unit members will respond better to leaders who are transparent and ready to listen and work with them to resolve their concerns. To assess whether the leaders' actions have addressed the concerns, another morale survey with adequate time interval should reveal the impact.
- Finally, leaders must be prepared to reflect and work on their own leadership behaviors.

some differences, personal morale in the U.S. Army units also correlated quite highly with the same factors. As for unit morale, unit cohesion and relationship with officers remained important, with confidence in company commander gaining greater importance over self-confidence. In essence, the study highlighted the factors of cohesion, leadership and self-confidence that were related to morale were also the same factors that underlie combat motivation as discussed in the last section. Finally, the mission-relevant factor in the Britt & Dickinson's (2006) model appears to be somewhat related to the ideology factor for combat motivation. Here, the importance of the mission, a clear purpose and achievable objectives parallel the specific reasons and beliefs one needs in order to go to war and risk one's life for a combat mission.

In a more recent study on U.S. military peacekeepers deployed to Kosovo, Maguen & Litz (2006) examined factors such as life stressors, cohesion and patriotism-nationalism on soldier's pre-deployment morale, and additional factors such as overseas-mission stressors and experience on post-deployment morale. They found that pre-deployment morale was most strongly related to unit cohesion, followed by other stressors in their daily lives during the survey, and a sense of patriotism-nationalism. Post-deployment morale was found to be significantly related to the deployment stressors, the peace-keeping experience they had, and post-deployment unit cohesion. The research by Maguen & Litz (2006) emphasize how stressors could negatively affect morale and highlight the positive relationships that unit cohesion and ideology (such as patriotism-nationalism beliefs and values) have with morale. Finally, the relationship between peace-keeping experience and post-deployment morale suggests that the soldiers' concept of morale includes an element of feeling good and satisfied (i.e., more than just the motivation to perform for the mission), and reinforces soldiers' the common use of the term, "morale".

SUMMARY

Soldiers' motivation and morale have long been recognized by commanders as the X-factor and force-multiplier in battles and operations to achieve mission success. It is, therefore, important for commanders to understand soldiers' motivation and morale so that they can exercise good leadership to inspire and motivate their soldiers to give their best and their lives for their nation and comrades if needed. Motivation arouses and directs attention and energy for behavior. It is manifested in the choice or direction, strength and persistence of the behavior. Motivation is dynamic and highly situation-dependent. There are a number of motivational theories. Two of the "content" theories (Maslow's hierarchy of needs and Herzberg's two-factor theory) and four of the "process" theories (reinforcement theory, expectancy theory, equity theory and self-determination theory) were described. Notwithstanding the inherent limitations in each, the theories still provide useful frameworks and models for commanders to use in motivating their soldiers.

Combat motivation is about why soldiers fight and what keeps them fighting in the face of threats. Soldiers need basic necessities such as food, drink, rest and sleep. When these basic needs for survival are not met, they can quickly drain the combat motivation of any soldier. However, an excess of these basic needs has minimal positive impact on combat motivation. History and studies have shown that cohesion, leadership, self-confidence, ideology and esprit de corps are important ingredients for combat motivation. "Fighting for my buddies" or primary group cohesion was the most frequently cited reason for combat motivation by soldiers in World War II and Operation Iraqi Freedom. Leadership and self-confidence are key in overcoming soldiers' fear and inertia to engage the adversary as well as to continue fighting. Ideology is important to bring soldiers to and keep them in the theater of operation but not sufficient to keep them fighting when bullets fly. Finally, esprit de corps provides additional motivation for soldiers to join the battle/operation, fight hard and stay on for the glory and reputation of their unit or organization.

Morale, compared to combat motivation, is a more commonly used term but its definition is less commonly agreed upon. Some equate it with combat motivation, some view it as an emotional state, while others consider it to be a psychological state that includes motivation toward mission success. The last definition appears to be gaining some consensus among researchers because it is the most useful and yet differentiated from motivation per se. As such, factors influencing morale tend to overlap greatly with the same factors cited for for combat motivation. Leaders' experience and research have supported the positive relationship between morale and outcomes such as soldiers' performance, psychological health and intention to stay in the unit or organization. These important links and the duty of leaders to build and sustain unit morale are emphasized in the leadership doctrines and guides developed by the U.S. Army, NATO, Canadian Forces and Australian Defence Force. In addition, the NATO RTO leadership guide specifically encouraged leaders to work with psychological support professionals to survey the morale of their soldiers and take actions to address their morale concerns.

Key Concepts

- Motivation
- Maslow's hierarchy of needs theory
- Herzberg's two-factor theory
- Reinforcement theory
- Expectancy theory
- Equity theory
- Self-determination theory
- Combat motivation
- Morale
- Cohesion
- Leadership
- Self-confidence
- Psychological operations
- Ideology
- Conscientious objectors
- Esprit de corps

Questions

- Are you currently facing any motivational challenges with your soldiers? Based on what you have read in this chapter, what is your understanding of their motivation and could you apply any of the ideas and principles to motivate them?
- If you were to conduct a morale survey in your unit today, which areas would you expect your soldiers to rate positively and what concerns would you expect to hear from them? How would you discuss and resolve these concerns?
- If your unit is to be deployed for a peace-keeping mission in eight weeks' time, what would you do to raise and sustain the combat motivation of your soldiers before and during deployment?

References

Australian Defence Force (2007). Australian Defence Doctrine Publication 00.6: Leadership in the Australian Defence Force. Canberra, ACT: Australian Defence Headquarters.

Baynes, J.C. (1967). Morale. New York: Praeger.

Belenky, G., Noy, S., & Solomon, Z. (1987). Battle stress, morale, cohesion, combat effectiveness, heroism, and psychiatric casualties: The Israeli experience. In G. Belenky (Ed.), *Contemporary studies in combat psychiatry.* (pp. 11–20). Westport, CT: Greenwood Press.

Britt, T.W., & Dickinson, J.M. (2006). Morale during military operations: A positive psychology approach. In T.W. Britt, C.A. Castro & A.B. Adler (Eds.), *Military life: The psychology of serving in peace and combat (Vol. 1): Military performance.* (pp. 157–184). Westport, CT: Praeger Security International.

Bond, T.C. (1976). The why of fragging. *American Journal of Psychiatry, 133(11)*, 1328–1331.

Canadian Forces (2005). Leadership in the Canadian Forces. Kingston, ON: Canadian Defense Academy.

Catignani, S. (2004). Motivating soldiers: The example of the Israeli Defense Forces. *Parameters, 34(3)*, 108–121.

Deci, E.L., & Ryan, R.M. (1985). *Intrinsic motivation and self-determination in human behavior.* New York: Plenum.

Deci, E.L. & Ryan, R.M. (2000). The "what" and "why" of goal pursuits: human needs and the self-determination of behavior. *Psychological Inquiry, 11*, pp. 227–268.

Franken, R.E. (2007). *Human motivation* (6th ed.). Belmont, CA: Wadsworth.

Gal, R. (1986a). *A portrait of the Israeli soldier.* London: Greenwood.

Gal, R. (1986b). Unit morale: From a theoretical puzzle to an empirical illustration — An Israeli example. *Journal of Applied Social Psychology, 6(6)*,

550.

Gal, R., & Manning, F.J. (1987). Morale and its components: A cross-national comparison. *Journal of Applied Social Psychology, 17(4)*, 369–391.

Griffith S.B. (1963). *The art of war*. Oxford University Press.

Kellet, A. (1984). *Combat motivation: The behavior of soldiers in battle*. London: Kluwer Publishing.

Maguen, S., & Litz, B.T. (2006). Predictors of morale in U.S. peacekeepers. *Journal of Applied Social Psychology, 36(4)*, 820–836.

Manning, F.J. (1991). Morale, cohesion and esprit de corps. In R. Gal. & D.A. Mangelsdorff (Eds.), *Handbook of military psychology* (pp. 453–470). Chichester, U.K.: John Wiley and Sons.

Marshall, S.L.A. (1947). *Men against fire*. NewYork: William Morrow.

Mitton, T.M. (2007). New considerations in morale for Canadian operational commanders. *The Canadian Army Journal, 10(3)*, 27–43.

Motowidlo, S.J., & Borman, W.C. (1977). Behaviorally anchored scales for measuring morale in military units. *Journal of Applied Psychology, 62*(2), 177–183.

Motowidlo, S.J., & Borman, W.C. (1978). Relationships between military morale, motivation, satisfaction, and unit effectiveness. *Journal of Applied Psychology, 63(1)*, 47–52.

NATO Research and Technology Organization (2008). *Stress and psychological support in modern military operations* (RTO Technical Report, TR-HFM-081). Retrieved Jan 2, 2011, from http://www.rta.nato.int/Pubs/RDP.asp?RDP=RTO-TR-HFM-081.

Oliver, L.W., Harman, J., Hoover, E., Hayes, S.M., & Pandhi, N.A. (1999). A quantitative integration of the military cohesion literature. *Military Psychology, 11(1)*, 57.

Petri, H.L., & Govern J.M. (2004). *Motivation: Theory, research and applications* (5th ed.). Belmont, CA: Wadsworth.

Rodrigues-Goulart, F. (2006). Combat motivation. *Military Review, 86(6)*, 93–96.

Siebold, G.L. (2006). Military group cohesion. In T.W. Britt, C.A. Castro & A.B. Adler (Eds.), *Military life: The psychology of serving in peace and combat (Vol. 1): Military performance.* (pp. 185–201). Westport, CT: Praeger Security International.

Stewart, N.K. (1989). South Atlantic conflict of 1982: A case study in cohesion. *Military Review, 69(4)*, 31–40.

Stouffler, S.A., Lumsdaine, A.A., Lumsdaine, M.H., Williams, R. M., Smith, M.B., Janis, I.L., Star, & Cottrell, L.S. (1949). *The American soldier*. Princeton, NJ: Princeton University Press.

U.S. Army (2006). Army leadership (FM6-22). Washington, D.C.: Headquarters, Department of the Army.

Van der Dennen, J. (2005). Combat motivation. *Peace Review, 17(1)*, 81–89.

Wong, L., Kolditz, T.A., Millen, R.A., & Potter, T.M. (2003). *Why they fight: Combat motivation in the Iraq War.* Carlisle, PA: Strategic Studies Institute.

Woods, S. & West, M. (2010). *The Psychology of work and organizations* (1st ed.). Belmont, CA: Wadsworth.

5

Military leaders — Born or Made?

Source: CyberPioneer, Ministry of Defence, Singapore

OVERVIEW

If the last three chapters focused on the operational and soldiering *context* of leadership, the next two chapters focus on the aspects of the *leader* in the leadership process. This chapter aims to provide an overview of the "trait" and the "behavioral" approaches to leadership by addressing the often asked question "Are leaders born or made?" Knowledge from the trait approach is relevant to military organizations because of the critical need to select and assign leaders to different roles and positions. Knowledge of the different kinds of "traits" such as personality, values and abilities can help leaders to appreciate how people are different from each other, which in turn enables the leaders to adapt to different people in the influence process. Similarly, knowledge from the behavioral approach is relevant for leaders' self-diagnosis of their current strengths and weaknesses or "behavioral competencies", which is a starting point for self-development. This chapter also introduces readers to how various military forces employ "competency models" to guide leadership development and training today, which is essentially related to the behavioral approach to leadership. Finally, readers will learn that the answer to the "born or made" question is not a simple one; that the research efforts attempting to answer this simple question have helped to explain why there is a lot more to leadership than common-sense!

EARLY TRAIT AND BEHAVIORAL APPROACHES: 1900s–1970s

The Early Search for "Leadership Qualities"

Throughout history, man has searched for certain qualities that characterize leaders. In his *Handbook of Leadership*, Bass (1990) observed that notions about leader qualities could be found in the writings of many ancient cultures, for example in the writings of ancient philosophers like Lao-tzu, Plato and his student Aristotle, and in the Christian bible.

Some of the earliest scientific theories of leadership began in the 18th century with a search for the set of traits that would "make a person a leader". In his "Great Man" thesis, Carlyle (1841) argued that history was shaped by the personal qualities of great leaders who were "born" rather than "made". Reflecting this view, Galton (1869), in trying to explain why the leaders of British society were more likely to be related to each other than chance alone might allow, concluded that it was the superior genes of the British ruling classes that determined their status in society. In one of the earliest empirical studies of leadership comparing leaders and non-leaders among school children, Terman (1904) reported following attributes as characterizing young leaders: They were verbally fluent, intelligent, low in emotionality, daring and lively, high in congeniality, and generally "good".

The U.S. Army introduced the first intelligence tests (the Army Alpha and Beta) designed for large-scale, "mass" testing of cognitive abilities during the First World War. Over 1.7 million men were administered such tests for selection or classification purposes from 1914 to1918. Starting from about the Second World War, the period from the 1940s to the 1960s witnessed a rising interest in "behaviorism", which held the view that psychological phenomena should be studied in terms of observable behaviors rather than intangible "constructs" like "traits". In 1948, Stogdill published in an influential review where two of his concluding remarks were interpreted by many to imply traits cannot predict leadership: "A person does not become a leader by virtue of the possession of some combination of traits" (p. 64), and, "persons who are leaders in one situation may not necessarily be leaders in other situations" (p. 65). This is even though Stogdill (1948) presented in his review evidence from over 15 empirical studies indicating that individuals holding leadership positions tended to be higher than their followers in terms of factors such as intelligence, scholarship, dependability in exercising responsibility, activity and social participation, and socio-economic status. Stodgill also reported evidence from another 10 studies that showed that leaders were on average higher on factors like sociability, (initiative, persistence, knowing how to get things done, self-confidence, alertness-to and insight-into situations, cooperativeness, popularity, adaptability, and verbal facility. In another review of the leadership literature, Mann (1959) concluded that there were positive but weak relationships between leadership and various personality characteristics. Specifically, he reported that leaders were higher in dominance, extraversion, interpersonal sensitivity, masculinity, conservatism, intelligence, and adjustment. These reviews resulted in many social psychology textbooks in the 1950s and 1960s emphasizing that traits cannot predict leadership. This led to a view that leadership could possibly consist of learned behaviors and that leaders were therefore more "made than born".

The Early Search for Leadership Behaviors

The most significant research studies conducted to determine the behaviors of effective versus ineffective leaders was initiated during the mid-1940s in what was called the "Ohio State Leadership Studies" (e.g., Shartle, Stogdill, & Campbell, 1949; Stogdill & Shartle, 1948). These studies started with careful observation to determine common patterns of leaders' behaviors across different leadership roles. Many of these early observations were conducted in the military. From these observational studies, questionnaires were then developed and administered to people in organizations who were asked to rate the frequency in which their leaders displayed each behavior. When these ratings were subject to a statistical technique called "factor analysis", two broad underlying factors were found to explain the different leader behaviors. These were: "consideration" and "initiation of structure" (Stogdill, Wherry, & Jaynes, 1953). "Consideration" referred to the extent which the leader showed concern and respect for the followers, looked out for their welfare, and expressed appreciation and support for their efforts. "Initiation of structure" referred to the degree to which the leader would define and organize his or her own role, and the roles of followers, is oriented toward achieving the outcomes, and establishes clearly-defined patterns of communication in the group. These factors were also reported in studies of "group dynamics" conducted at the University of Michigan, where the factors were labeled "concern for people" and "concern for

Consideration and initiation of structure are the most basic of leadership behaviors.

Source: CyberPioneer, Ministry of Defence, Singapore

production". "Task vs. People" were eventually incorporated into many "contingency" theories of leadership in the 1960s and 70s, including Fiedler's Contingency theory of leadership (1967), Hersey & Blanchard's (1969) Situational Leadership theory, and House's (1971) Path-goal theory. These theories are described further in Chapter 6 under "Styles".

In a recent review of the validity of consideration ("People") and initiating structure ("Task") leadership behavior dimensions, Judge, Piccolo and Illies (2004) note that these dimensions dominated leadership research in the late 1970s, after which they curiously became the "forgotten" constructs in the study of leadership. Some reasons for the diminished interest in these behavioral constructs stemmed from criticism that the Ohio State studies overly relied on survey data which were often obtained from the same observers or raters for both the leader's behavior and performance criteria (Schriesheim & Kerr, 1974). Judge et al. (2004) added, "House's (1977) theory of charismatic leadership shifted the focus of leadership research to higher end constructs. Further, with the emergence of implicit leadership theories, researchers started to question the internal validity of behavioral leadership survey measures in general, and measures of Consideration and Structure were no exception". In many ways, this reduced interest in a behavioral explanation of leadership reflected fundamental shifts in the paradigms of scientific psychology from behaviorism to situationism in the 1970s, followed by interactionism in the 1980s, and finally, cognitivism in the 1990s.

REVIVAL OF TRAIT APPROACH: 1970S TO PRESENT

Realizing that "Leadership Is not so Simple"

Applying innovative experimental research designs, Kenny and Zaccaro (1983) concurrently varied task and group composition factors to determine the extent to which leadership outcomes could be explained by task, group composition and individual (leader) factors. They found that a 49 per cent to 82 per cent variation in who would emerge as a leader in groups could be attributed to the leader's individual characteristics. A similar "rotational" design study by Zaccaro, Foti and Kenny (1991) found that individuals who were high in the "self-monitoring" trait tended to emerge as leaders. This is because high self-monitors are more concerned with the appropriateness of their actions in social situations and would adjust their behavior to fit different situations.

With advances in statistical techniques of aggregating results across numerous studies, a Lord, De Vader, and Alliger (1986) conducted a "meta-analysis" that corrected the reported correlations in past studies for various statistical artifacts (e.g., sampling error, restriction of range). They reported the estimated correlation between various traits with leadership across numerous studies as follows: r=.50 for cognitive ability or "general intelligence", r=.24 for "adjustment" or emotional stability, r=.13 for "dominance"; and r=.26 for "extraversion". They concluded, "personality traits are associated with leadership *perceptions* (italics added) to a higher degree and more consistently than the popular literature indicates".

One important lesson learned in attempting to answer the "born or made" question was that "leadership" itself was not a simple, unitary phenomenon – that, in trying to determine traits that could predict leadership, it was important to distinguish between perceptions of leadership or "leader emergence", from actual "leadership effectiveness" or "performance". Hogan, Curphy and Hogan (1994) noted, "research on leadership emergence identifies the factors associated with someone being perceived as leaderlike" (p. 496). In a review of personality and leadership, Judge et al. (2002) commented: "leader emergence refers to whether (or to what degree) an individual is viewed as a leader by others, who typically have only limited information about that individual's performance. In contrast to being perceived as a leader, leadership effectiveness refers to a leader's performance in influencing and guiding the activities of his or her unit toward achievement of its goals" (p. 767). Today, it is generally recognized that traits are generally better predictors of leader emergence than of leadership effectiveness (Northouse, 2007).

Enhanced Understanding of "Traits"

Another lesson learned in the search for leader traits was the importance of having a common language to talk or think about "traits" (e.g., the "Big Five" Factors) — many of the early studies used different terms to refer to similar traits. The 1980s witnessed a renewed interest in the general study of traits or individual differences in psychology, and by the 1990s, a consensus emerged in the scientific community that traits could be broadly categorized into the ability and non-ability categories (also loosely called "can do" and "will do" traits; see Figure 5-1).

Traits are scientific "abstractions" or "constructs" that are inferred from observations that people are different from one another in consistent and systematic ways. Among the ability traits, cognitive ability is known to be highly predictive of job and leadership performance; and, different cognitive abilities are recognized to correlate highly to form a "general factor" called "general intelligence", with the main sub-factors being quantitative and verbal abilities (see Carroll, 1993). Non-ability traits include individual differences in personality, values and interests. Today, personality is seen as most predictive of work performance and is best organized in terms of five relatively independent factors or dimensions, which in turn comprise several narrow "facets". Both genetic and environmental factors influence the development of ability and personality traits; however, some traits are primarily shaped by genetic factors. For example, it is estimated that about 60% of the variability in general intelligence or cognitive ability is due to genetic factors (Vernon, 1979). Zuckerman (1991) reported that the main personality traits of extraversion, neuroticism, and psychoticism are also primarily shaped by genetic factors.

Personality refers to relatively stable characteristics that describe a person, or "what makes the person unique or different from others in terms of their behavior, feelings and thinking". The term

FIGURE 5-1. The language of "traits" today

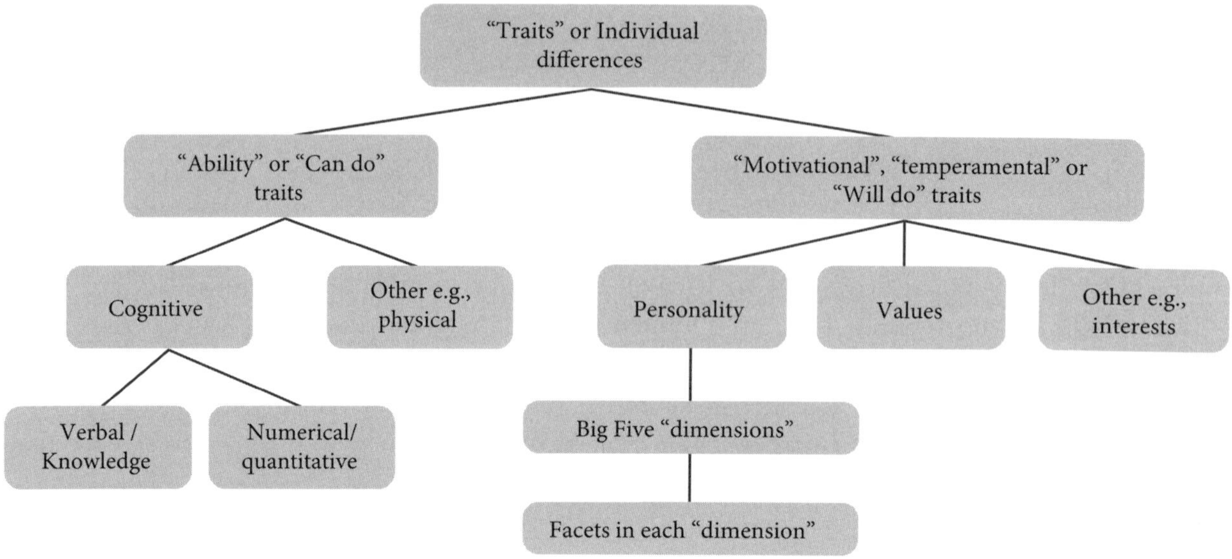

personality does not refer to transient moods or "states of mind" or attitudes; neither does it include abilities or "intelligences" or aptitudes or "can do traits". In the 1980s, several different groups of researchers independently concluded that almost all personality traits could be represented in terms of five broad trait factors. These were commonly labeled extraversion, neuroticism, conscientiousness, openness to experience, and agreeableness (Costa & McCrae, 1992; Digman, 1990; Goldberg, 1993). These traits are collectively referred to as the Five Factor Model (FFM) of personality or the Big Five. They originated from studies of trait-descriptive adjectives drawn from the lexicon.

Today, the Big Five model is the most widely used classification system for personality traits because it represents a unified and parsimonious theoretical framework for personality (Digman, 1990; Digman, 1997; Wiggins & Trapnell, 1997). Interestingly, some of the earliest ideas about the "Big Five" personality factors were first reported by U.S. Air Force researchers (Tupes & Christal, 1961) at a time when most researchers had in fact lost interest in the study of personality. A recent meta-analysis by Judge et al. (2002) reported an overall multiple correlation of .48 between Big Five personality factors with different leadership criteria (i.e., emergence vs. effectiveness; refer to Chapter 1 for discussion on this issue), which was higher than the overall correlation reported between general intelligence and leadership at .21 (Judge et al., 2004; see also Table 5-1). Judge et al. (2002) concluded that extraversion was the strongest predictor of leadership followed by conscientiousness, openness to experience and emotional stability. They also concluded that extraversion and conscientiousness both predicted emergence better than effectiveness criteria. They also noted that agreeableness was only weakly related to leadership outcomes, especially leader emergence.

TABLE 5-1. The Five Factor Model (FFM) of personality.

Big Five Factor	Description	Main "Facets" of this personality factor	Correlation with Leadership criteria	
			Emergence	Effectiveness
Neuroticism (opposite of "Emotional Stability")	The degree to which one is emotional, insecure, anxious and depressed versus self-confident and calm)	**Volatility**: Anger and irritability **Withdrawal**: Depression & worry	−.24	−.22
Extraversion (opposite of "introversion")	The extent to which one is gregarious, sociable, and assertive versus timid, reserved and quiet.	**Enthusiasm**: Active, friendly **Assertiveness**: Dominance	+.33	+.24
Openness to Experience (also called "Intellectance")	The degree to which a person is curious and creative with wide interests versus having only narrow or limited interests.	**Intellect**: Quick thinking **Openness**: Curious, artistic	+.24	+.24
Agreeableness	The degree to which one is cooperative, warm, caring and nice to others versus cold, antagonistic and disagreeable.	**Compassion:** Care/Feel for others **Politeness:** Not hostile or imposing	+.05	+.21
Conscientiousness	The extent to which one is organized, hardworking, dependable, rule-abiding and persevering versus lazy, disorganized, and unreliable.	**Industriousness:** Achievement **Orderliness:** Like rules, structure	+.33	+.16

Note: Average correlations with leadership criteria are as reported in Judge et al. (2002). "Facets" are as reported in DeYoung et al. (2007).

In a recent review of the literature on leadership traits, Zaccaro, Kemp, & Bader (2004) defined "leader traits" as "relatively stable and coherent integrations of personal characteristics that foster a consistent pattern of leadership performance across a variety of group and organizational situations." These characteristics reflect a range of stable individual differences, including personality, temperament, motives, cognitive abilities, skills, and expertise. They concluded that several classes of traits predict both leader emergence and performance, including:

- Cognitive capacities, e.g., general intelligence and creative thinking capacities;
- Personality, including the "Big Five" personality factors;
- Motives and needs, e.g., the need for power, achievement and the motivation to lead;
- Social capacities, including self-monitoring, social and emotional intelligence;
- Problem-solving skills, including problem construction, solution generation and meta-cognition; and,
- Tacit knowledge.

Understanding Why and How Traits Relate to Leadership

Traits matter because of perceptions and expectations of leaders. Knowing that certain traits may predict or correlate with leadership is one thing; understanding how and why traits can or should relate to leadership is another. In discussing their finding of relatively strong correlations between leadership and personality traits, Judge (2002) noted that behavioral scientists still only had a relatively poor idea of "not only which traits are relevant, but why". They speculated that one explanation could be in the expectations or stereotypes that followers have of leaders: for example, it could be that people tend to expect leaders to be more sociable and assertive, and that is why these traits best predict leadership. But, they also acknowledged that the correlations found between traits and leadership criteria do not provide a causal understanding of why traits matter for leadership. In another meta-analytic study, Judge et al. (2004) also suggested that the link between intelligence and leadership could be a perceptual one: "Because the emergence of leadership is in part a product of impression or image management (Chemers, 2001; Gardner & Avolio, 1998), appearing smart may be more important than being smart (Rubin et al., 2002)" (p. 543). Indeed, they confirmed this notion when they found that perceptual measures of intelligence and leadership produced higher correlations than objective measures of these constructs.

Traits can act as a resource to buffer leaders under different stress. In 1986, Fred Fiedler proposed a Cognitive Resource Theory of leadership, suggesting that both intelligence and experience were important factors affecting leadership effectiveness but in different ways. In this theory, leader intelligence and experience was seen as irrelevant for simple tasks. However, leader intelligence or cognitive ability affected leadership performance most in low stress situations when the leader's role is mainly a directive one, while experience mattered more than intelligence in high stress situations because having experience would help the leader to make better quality decisions. This theory received generally good experimental support. A meta-analysis examining the relationship between intelligence and leadership by Judge et al. (2004) also found empirical evidence supporting the idea that intelligence affects leadership performance most in low stress situations when the leader's role is mainly a directive one. Taken as a whole, this suggests that the understanding of how traits relate to leadership should go beyond an assumption that the mere presence of a trait "causes" leadership. Instead, it is important to appreciate how traits affect leadership performance through mechanisms like thinking and decision making, and how these effects also depend on other factors like the leader's experience level and the level of stress in the situation.

A motivational-developmental theory explaining how traits may relate to leadership outcomes. Chan & Drasgow (2001) proposed a theory that suggested an indirect way in which traits may relate to leadership performance, through motivational and developmental processes. Breaking away from the simplistic assumption that there existed "leadership traits or genes" that act like "on/off" switches for leadership; they suggested that traits like personality and values predisposed certain individuals to be more motivated to lead, and that this motivation to lead (MTL) caused them to seek and to participate in leadership or leadership learning opportunities from which they could acquire the skills, abilities, styles and experience (or tacit knowledge) for leading. The latter in turn represented the total set of "resources" that a leader could bring to a leadership situation (including task, follower, and contextual conditions) to affect outcomes (i.e., "leadership performance" or "expectations"; see Figure 5-2). In their empirical research involving both Singapore military soldiers and students, and U.S. students, they demonstrated how "distal traits" such as personality and social values affected MTL, and how MTL itself could predict "leader emergence" criteria. In essence, the theoretical framework by Chan and Drasgow (2001) represents a significant shift from the "Great Man theory" assumption that leaders are either born or made, to one that stresses the role of traits in affecting leader emergence and performance indirectly through leadership development activities and experiences which affect the acquisition of leadership skills and competencies. This framework therefore bridges the trait and behavioral approaches to leadership while also accounting for the "lessons learned" from trait approach (e.g., the importance of distinguishing different leadership criteria; the importance of situational contingencies and the leader's resources in affecting leadership performance).

FIGURE 5-2. Chan & Drasgow's (2001) theoretical framework linking traits with various leadership outcomes.

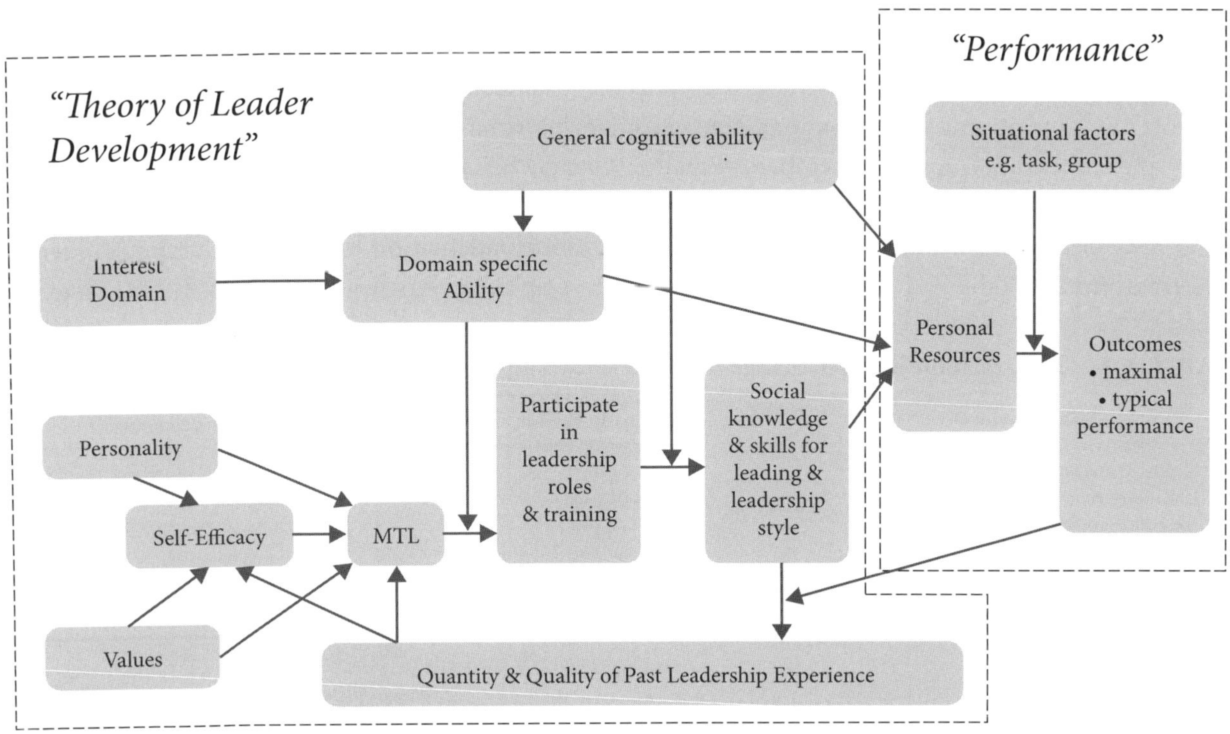

Note: MTL refers to "Motivation to lead".

Source: Chan & Drasgow (2001).

WHAT WE KNOW ABOUT LEADERSHIP BEHAVIORS TODAY

Task, Relations and Change Behaviors

Throughout the 1980s and 1990s, when most other leadership researchers focused on cognitive or transformational/charismatic approaches to leadership, Yukl (1999) continued to study leadership behaviors in the tradition of the Ohio State studies. In 1999, Yukl noted, "Most of the well-known theories of leadership effectiveness were initially formulated using a two-factor conception of leader behavior or traits. Examples include task-oriented versus relations-oriented leadership, autocratic versus participative leadership, leadership versus management, transformational versus transactional leadership, and charismatic versus non-charismatic leadership. These dichotomies provide some insights, but they also oversimplify a complex phenomenon and encourage stereotyping of individual leaders". He reported a study where he combined the items of various leadership questionnaires, and where factor analysis of the data revealed three broad leadership behaviors, namely, Task, Relationship and Change (TRC) behaviors. Yukl (1999) argued that these were three exclusive "meta-categories" of leadership behaviors that differed in "the primary objective of the behavior". In Yukl's view, "Task behaviors" are those that aim to achieve a high level of efficiency in the use of resources (including personnel), and high degree of reliability of operations, products, and services. In contrast, "relations behaviors" are those that leaders may use to achieve a feeling of commitment to the group, or organization and its mission. Relations behaviors are important for building a high level of mutual trust and cooperation among members. Finally, "change behaviors" are those that aim to achieve innovations and improvements in terms of work processes, "products or services", and the organization. They are essential for facilitating the group or organization's adaptation to changing circumstances in the environment. Table 5-2 summarizes the key leadership behaviors organized by the TRC factors.

TABLE 5-2. Task, relationship and change-oriented leadership behaviors.

Task Behaviors	Relations Behaviors	Change Behaviors
• Short-Term Planning • Clarifying Responsibilities • Monitoring Operations and Performance	• Delegating • Supporting • Developing • Recognizing • Consulting • Empowering	• External or "environmental scanning" • Envisioning Change • Encouraging Innovative Thinking • Taking Personal Risks

Source: Yukl (1999).

By now, readers will notice that research on leadership is affected by the changing research paradigms in psychology. Although "discovered" in the 1940s, interest in initiating structure and consideration as the broad dimensions of leadership behavior was somehow eclipsed by other behavioral models, especially the transformational-transactional and charismatic leadership models that dominated the literature in the 1980s and 90s, and, by other views of leadership, e.g., the information processing approach. It was only in 2004 that a meta-analytic review by Judge et al. (2004) revived interest in these constructs by reporting "average" correlations of .48 and .29 between consideration and initiating structure respectively with leadership outcomes across hundreds of studies. More importantly, they showed that consideration or "concern for people" was more strongly related to follower satisfaction and motivation and ratings of leader effectiveness, while initiating structure was slightly more strongly related to measures of the leader's job performance and also to group or organizational performance. To put it simply, leaders who show more behaviors reflecting "concern for people" will have followers who are more motivated and satisfied with the job and the leader, while those who demonstrate more concern for task or production yield stronger group or organizational performance. Because these factors are independent, it is possible for leaders to be high in both kinds of behaviors; i.e., effective leadership is not about task *or* people behaviors but demonstrating both kinds of behaviors appropriately. Overall, this meta-analysis provided strong support for the validity of Initiating Structure and Consideration as behavioral constructs in leadership research and theory.

"Cognitive Skills" Needed for Military Leadership

With the proliferation of personal computing in the 1990s, military operations become increasingly "digitized" and this significantly increased the cognitive demands on all soldiers and especially military leaders on the battlefield. Recognizing these increased cognitive demands, the U.S. Army Research Institute supported various research programs that looked into the mental skills that leaders would need to solve ambiguous problems throughout the 1990s.

Military leaders need a high degree of cognitive skills to operate in the 21st century context alongside task, people and change-related skills.
Source: CyberPioneer, Ministry of Defence, Singapore

After several years of military research, Mumford, Zaccaro, Harding, Jacobs, and Fleishman (2000) argued that in a complex operating environment, leaders need to hone their cognitive skills in "problem definition", gathering information, formulating ideas and constructing plans to solve problems. They proposed a skills-based model of leader performance that suggested that leadership can be understood in terms of three key types of cognitive skills, including (1) complex problem-solving skills, (2) solution construction skills, and (3) social judgment skills. They emphasized that these skills were learned via the interaction of traits and actual experience, and that these skills should be complemented with relevant expert knowledge for effective performance.

"Competencies" Required for Military Leadership

The popularity of "competency modeling" grew in the 1990s in response to an accelerated pace of change in the complexity of work. The idea of "competencies" was first proposed by McClelland (1973) who suggested that competencies should be adopted as a more useful approach to aptitude measurement as part of personnel selection rather than techniques such as intelligence tests which seemed too far removed from the practical realities of work. Unlike the traditional job analysis approach where each job is analyzed and described in terms of often unique behavioral requirements, competency modeling begins by identifying a common set of "competencies" for an organization in which all jobs are referenced. Competencies are typically identified in a manner that is relevant to the organization's current and future mission and operations, and sometimes even aligned to the organization's own "competencies" (note: individual level competencies are not identical to organizational competencies).

The purpose of competency modeling is to provide a common framework or language for describing different leadership roles across organizations, so as to allow for greater flexibility for the adaptation of such jobs or roles to changing organizational needs. Today, many organizations use competencies models to guide their design of their human resource systems including selection, training and development, performance management, etc. Some have raised concerns that the practice of competency modeling has proliferated much faster than the science and methods for creating valid models. Montebello (2001), for example, commented that "it is not uncommon to find organizations copying the competency models from other companies simply because they 'like what they see' or 'the competencies appear relevant and applicable" (p. 1034).

Figure 5-3 presents examples of how two different military forces use the idea of competencies in their leadership doctrines. In the U.S. Army's BE-KNOW-DO leadership framework which describes its ideal of leaders of "character and competence", it is not surprising that competencies and values form the "common language" for describing leadership. In its latest leadership doctrine, the U.S. Army (2006) explains that its core leader competencies "emphasize the roles, functions, and activities of what leaders do" and stem directly from the Army definition of leadership. Its competencies are presented alongside a list of "attributes" like "character", "presence" and "intellectual capacity" as part of its broader "Leadership Requirements Model". Similarly, the Singapore military has its own leadership competency model which is part of a broader doctrine framework called "SAF Leadership 24-7"

FIGURE 5-3. Example of two military leadership competency models.

U.S. Army Core Leadership Competencies "What an Army Leader Does" (FM6-22, 2006)	Singapore Armed Forces Leadership Competencies (from Chan et al., 2005)
Leads • Leads others • Extends influence beyond the chain of command • Leads by example • Communicates Develops • Creates a positive environment • Prepares self • Develops others Achieves • Gets results	Conceptual thinking competency • Critical thinking • Creative thinking • Ethical reasoning Mission competency • Planning • Implementing • Decision-making Social competency • Communicating to influence • Interpersonal effectiveness Developmental competency • Developing People • Developing Teams • Improving organization Self meta-competency • Self-awareness • Self-management • Personal Mastery

Source: U.S. Army, 2006; Chan et al., 2005

(see Figure 1-2) that also includes non-behavioral competency domains like "values" and "styles". The Singapore military's leadership competency model organizes 14 behavioral skills into five broad "competencies", one of which — "self" — is called a "meta-competency" because the skills involved (e.g., personal mastery, self awareness and self management) are deemed important for developing other skills in the competency model (see Chan et al., 2005, for more on the Singapore military's leadership framework and competency model).

APPLYING THE SCIENCE OF LEADERSHIP TRAITS AND BEHAVIORS

Traits and Military Leader Selection

Just as the scientific study of leadership only started in the late 19th century, it was also during that period with the rise of the industrial age that scientific personnel selection emerged in organizations. The idea of systematic selection is based on the idea that relatively stable individual differences (or traits) exist and can be reliably measured, and that jobs, tasks and roles also differ in systematic ways in terms of the human abilities and qualities needed to perform them effectively. If these are

A sweating candidate concentrates during British army officer selection tests in 1942. Today, most forces continue to utilize psychological tests for leader selection.

Source: Fred Ramage/Hulton Archive/Getty Images

possible, then organizations may measure certain individual traits to predict future performance in various jobs and roles (Cascio, 1991; Dunnette, 1966). Personnel selection refers to the science of measuring individual differences or traits that have been determined through validation to predict future job/role performance, and to use this information to help companies or organizations determine which job applicant or candidate is most qualified for the job as part of hiring or "advancement" decisions. Classification refers to the use of the same information from a group of people to establish the most optimal "fit" of the group of individuals to a range of different jobs.

"Selection procedures" (also called tests, predictors, and assessments) refer to the techniques and methods for measuring various individual characteristics which form the basis for making personnel decisions like hiring, training, job assignment, promotion, etc. These include paper and pencil tests, inventories (e.g., personality, interest), computer-administered tests, assessment center evaluations (sometimes called "situational tests"), biographical data forms, interviews, educational and/or experience requirements, background investigations, reference checks, physical requirements (e.g., height or weight) and ability tests, etc. Over 100 years of the science and practice of personnel selection have resulted in meta-analytic studies providing highly stable estimates of the validities of different selection methods in predicting general job performance (e.g., Schmidt & Hunter, 1998). Among the most predictive selection techniques are cognitive ability or IQ tests, work sample tests (also called "situational tests" or assessment centers), structured interviews and peer ratings.

With large numbers of personnel and a diverse range of vocations and roles, each requiring certain traits or aptitudes, military forces have in the last century played a leading role in the design and development of new techniques for personnel leadership selection. For example, the first intelligence tests designed for large scale testing — the Army Alpha and the Army Beta — were designed by psychologists and administered to over 1.5 million army recruits from 1917 to 1918 as part of the effort to screen for basic cognitive abilities needed for military service, to classify individuals into different military jobs, and to select for leadership potential. From the 1920s to the 1940s, the German military developed and employed the first "assessment centers" for the selection of officers.

They placed candidates in highly naturalistic settings and employed a range of observers including psychologists, physicians and actual military officers to make somewhat clinical judgments of the leadership potential of the candidates in a holistic way over two to three days. These highly systematic and realistic techniques for leadership selection were studied by the British and adapted for use in "War Office Selection Boards" starting from 1942. The U.S. military also started using Assessment Centers for the selection of special personnel such as spies in its Office of Strategic Services (OSS) in late 1943. It was only in the 1950s that non-military organizations such as AT&T started to employ the assessment center methodology for selection. Today, assessment centers typically involve a group of participants taking part in a variety of realistic exercises or work samples or simulations that are observed by a group of trained assessors who evaluate them on a set of predetermined behavioral dimensions. These behavioral observations are "integrated" with findings from IQ and personality tests before a final recommendation is typically made whether to select a person for a certain kind of position. The idea of using multiple assessors to observe behaviors over multiple tasks and the use of multiple measurement methods is to increase the reliability and validity of the assessment. The use of work sample or "situational test" exercises increases the realism of the process. Another notable military contribution to the science of leadership selection is the development of peer appraisals (e.g., Hollander, 1954, 1965) which involves getting members of a group to evaluate fellow group members on specific traits and behaviors. The advantage of using peer ratings is that it involves multiple raters which increases the reliability of the assessment, and, peers may be able to observe and rate each other in everyday settings that superiors may not have access to. A study conducted with the over 125 platoons from the Israeli Defense Force (Amir, Kovarsky & Sharan, 1970) showed that peer appraisals collected during the early stages of military training could predict their soldiers' promotion to leadership ranks with a high degree of validity (note: correlations around .6 to .7).

Leader selection is a central issue for many modern military forces in that it affects the quality of individuals that they assign into leadership ranks (e.g., officers, non-commissioned officers). Military leadership selection has both immediate and long-term implications. In most immediate terms, selection affects a person's successful completion of the military leadership (e.g., officer) training program. On a longer term basis, those responsible for military leadership selection realize that a newly commissioned entry-level officer may one day become a General responsible for the lives of hundreds of thousands of troops. Other factors that affect the selection process include both the changing nature of military missions and systems which present new challenges for military leadership, and, the changing nature of society which itself affects the quality of new entrants into the military system, and also the competition for talent in the job market. While each military force's leadership selection decisions are rooted in their national culture and traditions, it is important to recognize that there exists a science of personnel selection in the field of industrial-organizational psychology with well-established methodologies, principles and standards, with systematic and structured techniques to select people for jobs that minimize clinical judgment and that make predictions based on more reliable and valid measurements.

Military Leadership Development

Until the 1990s, while there was a flourishing and ever-growing practice and business of leadership and management training, there was hardly a coherent science of "leadership development". In 1992, Conger published *Learning to Lead* in which he evaluated different forms of leadership or management training that were commonly implemented by organizations, including formal courses, adventure training, job assignments, etc. From this study, Conger concluded that the formal leadership training typically conducted in classroom environments at best helped to build awareness of leadership issues. His research showed that the most critical factors that resulted in actual leadership development in terms of changes were time and on-the-job experience, and the individual learner's motivation, ability and opportunity to learn, reflect, and to "make the change". Conger (1996) argued that effective leadership training required a combination of different methods including skill building, personal growth (including self-awareness), feedback and conceptual awareness.

In 1998, the non-profit organization Center for Creative Leadership published a *Handbook of Leadership Development* (McCauley, Moxley & Van Velsor, 1999) which articulated three factors shown in its research to be "keys" to leadership development. These factors were labeled:

- **Assessment** (of one's present strengths and developmental needs);
- **Challenge** (which refers to difficult tasks or assignments that call for new skills and perspectives and that cause the person to question his or her own established ways of thinking and acting); and,
- **Support** (which refers to a supportive, learning environment that builds a person's confidence on leading others).

At the turn of the millennium, the U.S. Army Research Institute commissioned several research efforts in an attempt to understand the principles and processes involved in developing leadership. One effort that was led by former Israeli military psychologists Popper, Amit, Gal, Mishkal-Sinai and Lisak (2007), reviewed the scientific literature on leadership, learning and human development and proposed that "in the broadest terms", leadership development was a multiplicative function of three sets of factors: Leadership Potential (P), Motivation to Lead (M) and the Developmental process (D). In this model, the term "leadership potential" referred to various leadership traits or "capacities" that are "a necessary but not sufficient condition (for leadership)". Another effort by Day and Halpin (2002) examined the best leadership development practices of the best commercial companies, for example in the use of coaching, mentoring, 360-degree feedback, etc., and concluded that "the most important principle in successful leadership development efforts is the presence of an influential champion", and, that "effective leadership development appears to be a function of the interdependence of the various practices rather than a collection of independent programs" (p. vii). Day and Halpin (2002) advised that rather than trying to adopt different new "best practices", organizations should instead focus on consistently implementing a set of leadership development practices to yield the best outcomes.

Leadership is best developed through real or simulated experiences with elements of assessment, challenge and support.

Source: CyberPioneer, Ministry of Defence, Singapore

With a greater appreciation of these principles, modern military forces are beginning to stress the importance of having a systematic approach to leadership development. For the Singapore Armed Forces, being "systematic" has meant introducing a consistent set of leadership frameworks and models to guide the many processes required for consistent development of its leaders throughout their military careers. These include the Singapore military's Leadership Competency Model (see Figure 5-1) and the SAF Leadership Framework (Figure 1-2) which itself includes four different "facets" or "domains" of leader development, including the "self", "values", "behavioral competency", and "styles" (see Chan et al., 2005; Lim, et al., 2005).

The U.S. Army has also continued to refine its competency-based approach to leadership development with the publication of its latest leadership doctrine manual in 2006. The developers of the 2006 U.S. Army Leadership Requirements Model noted, "Competencies, if well-defined and comprehensive, should provide individuals and organizational processes with the roadmap that identifies successful performance of leadership duties and responsibilities. . . . Competencies are a means to define and communicate leadership requirements in organizationally relevant terms. The competencies that are essential, or core, to all Army leaders will focus their developmental efforts on attaining Future Force capabilities. . . . an Army leadership framework that incorporates core competencies provides a common denominator for leader development" (p. 2, Horey, Falleson, Morath, Cronin, Cassella, & Franks, 2004).

SUMMARY

This chapter has attempted to relate the cumulative growth in the social scientific knowledge of the trait and behavioral approaches to leadership – each taking an opposite view of the classic question whether "leaders are born or made". First, we related how the trait approach, beginning with "Great Man theory" assumed somewhat simplistically that there were traits that leaders inherently possessed. This early approach failed partly because of a lack of a common language and framework for "traits", and, because we started to realize the importance of distinguishing between different leadership criteria, e.g., leader emergence and leadership effectiveness. Failure of the early trait approach gave rise to the behavioral approach in the 1940s and 1950s, which assumed that leadership could be learned in terms of various behaviors. Several research programs identified task (or "initiating structure") and people (or consideration) behaviors, and, which subsequently became the basis for the situational and contingency approaches to leadership. With enhanced understanding of the nature of individual differences and advances in research methodologies, the trait approach was "revived" in the 1980s. Today, we know that certain cognitive capacities, personality traits, motives and needs, social capacities, problem solving skills and tacit knowledge are especially able to predict leadership emergence, and to a weaker extent, leadership effectiveness or performance. Today, social scientists are developing and testing theories to explain why and how traits should be related to leadership outcomes. In the terms of behaviors, social scientists today recognize task, relations and change (TRC) as three of the broadest categories of leadership behaviors. While TRC behaviors are both observable and demonstrable, social scientists also recognize that leaders need to possess certain cognitive skills including complex problem-solving, solution construction and social judgment skills to deal with the increasingly complex operating environment.

In presenting the advancement of knowledge in the trait and behavioral aspects over the last century, we have attempted to show how the social scientific knowledge of leadership is shaped by changing paradigms and advances in theory and research methodology. We also tried to highlight the military's interest in gaining a better understanding of leadership traits and behaviors. We hope that readers now better appreciate why the answer to the "born or made" question is not a simple one; that the research efforts attempting to answer this simple question have helped us to understand why there is a lot more to leadership than common-sense intuition!

Today, many military organizations employ the knowledge gained from the trait and behavioral approaches to leadership in the design of their systems for leadership selection and development. Because intelligence and personality traits show a high level of temporal stability and are predictive of leadership outcomes — especially "emergence", these individual differences are widely used as factors in military leadership selection procedures today. Increasingly, military forces are also investing in systematic efforts to enhance leadership development throughout their organizations. Being systematic involves adopting leadership frameworks and models — usually based on scientific principles — to guide the organization-wide employment of certain leadership development "best practices".

Key Concepts

- Traits (and/or Individual Differences)
- Personality (or "will do" traits)
- Ability (or "can do" traits)
- "Big Five" or Five Factor Model of personality
- Initiating Structure
- Consideration
- Task-Relations-Change (TRC) behaviors
- Cognitive skills
- Meta-analysis
- Competency modeling
- Selection (vs. Classification)
- Leadership development
- Leadership potential
- Assessment, Challenge and Support

Questions

- Does your military organization place more emphasis on leadership selection or leadership development? What are the dominant assumptions or factors that shape your military's emphasis on leader selection versus development?
- Take a look at your military's leadership doctrine. To what extent does the doctrine spell out trait and/or behavioral competency requirements expected of military leaders? Are there any requirements that fall outside of what are normally called traits or behaviors? What are these requirements (e.g., values? beliefs? philosophies)?
- In your view, is it important when discussing "leadership" to differentiate between criteria like emergence and effectiveness? Can you think of military leaders who had leader-like traits who would emerge as leaders but were not necessarily effective as leaders? Conversely, are there military leaders you know who were highly effective but did not possess leader-like traits?

References

Amir, Y., Kovarsky, Y., & Sharan, S. (1970). Peer nominations as predictors of multistage promotions in a ramified organization. *Journal of Applied Psychology, 54*, 462–469.

Bass, B.M. (1990). *The Bass handbook of leadership: Theory, research, and managerial applications.* NY: Free Press.

Carlyle, T. (1907; original 1841). *On heroes, hero-worship, and the heroic in history.* Boston: Houghton Mifflin.

Carroll, J.B. (1993). *Human cognitive abilities: A survey of factor-analytical studies.* New York: Cambridge University Press.

Cascio, W.F. (1991). *Applied psychology in personnel management (4th ed.).* Englewood Cliffs, NJ: Prentice-Hall International.

Chan, K., & Drasgow, F. (2001). Toward a theory of individual differences and leadership: Understanding the motivation to lead. *Journal of Applied Psychology, 86*, 481–498.

Chan, K., Singh, S., Ramaya, R., & Lim, K.H. (2005). *Spirit & System: Leadership Development for a Third Generation SAF [4th Pointer Monograph]*. Singapore: SAFTI Military Institute.

Chemers, M.M. (2001). Leadership effectiveness: An integrative review. In M.A. Hogg & R.S. Tindale (Eds.), *Blackwell handbook of social psychology: Group processes* (pp. 376–399). Malden, MA: Blackwell.

Costa, P.T., Jr., & McCrae, R.R. (1989). *The NEO-PI/NEO-FFI manual supplement. Odessa*, FL: Psychological Assessment Resources.

Costa, P.T., & McCrae, R.R. (1992). *NEO PI-R. Professional manual*. Odessa, FL: Psychological Assessment Resources, Inc.

Conger, J.A. (1992) *Learning to Lead*. San Francisco: Jossey-Bass.

Conger, J.A. (1996, winter). Can We Really Train Leadership? *Strategy & Business, 2,* 52–65.

Day, D.V., & Halpin, S.M. (2001). *Leadership Development: A Review of Industry Best Practices [Technical Report 1111]*. Fort Leavenworth, KS: U.S. Army Research Institute Fort Leavenworth Research Unit.

DeYoung, C.G., Quilty, L.C., & Peterson, J.B. (2007). Between facets and domains: 10 Aspects of the Big Five. *Journal of Personality and Social Psychology, 93*, 880–896.

Digman, J.M. (1990). Personality structure: Emergence of the five-factor model. *Annual Review of Psychology, 41,* 417–440.

Digman, J.M. (1997). Higher-order factors of the Big Five. *Journal of Personality and Social Psychology, 73,* 1246-1256.

Dunnette, M.D. (1966). *Personnel selection and placement*. London: Tavistock.

Fiedler, F.E. (1967). *A Theory of Leadership Effectiveness*. New York: McGraw-Hill.

Fiedler, F.E. (1986). The contribution of cognitive resources ot leadership performance. In L. Berkowitz (ed.), Advances in experimental social psychology. NY: Academic Press

Galton, F. (1869/1892/1962). *Hereditary Genius: An Inquiry into its Laws and Consequences*. Macmillan/Fontana, London.

Gardner, W.L., & Avolio, B.J. (1998). The charismatic relationship: A dramaturgical perspective. *Academy of Management Review, 23*, 32–58.

Goldberg, L.R. (1993). The structure of phenotypic personality traits. *American Psychologist, 48*, 26–34.

Hersey, P., & Blanchard, K.H. (1969). Life cycle theory of leadership. *Training and Development Journal, 23(5),* 26–34.

Hogan, R., Curphy, G.J., & Hogan, J. (1994). What we know about leadership: Effectiveness and personality. *American Psychologist, 49*, 493–504.

Hollander, E. P. (1954). Peer nominations on leadership as a predictor of the pass-fail criterion in naval air training. *Journal of Applied Psychology 38*, 150–153.

Hollander, E. P. (1965). Validity of peer nominations in predicting a distant performance criterion. *Journal of Applied Psychology, 49*, 434–438.

Horey, J., Falleson, J.J., Morath, R., Cronin, B., Cassella, R., Franks, W., Jr., et al. (2004, July). *Competency based future leadership requirements* (Tech. Rep. No. 1148). Arlington, VA: United States Army Research Institute for the Behavioral and Social Sciences.

House, R.J. (1971). A path-goal theory of leader effectiveness. *Administrative Science Quarterly, 16,* 321–339.

House, R.J. (1977). A 1976 theory of charismatic leadership. In J.G. Hunt.Sc L.L. Larson (Eds.), *Leadership: The cutting edge: 189–207.* Carbondale, IL: Southern Illinois University Press.

Hunter, J.E. (1986). Cognitive ability, cognitive aptitudes, job knowledge, and job performance. *Journal of Vocational Behaviour, 29,* 340–362.

Hunter, J. E., & Hunter, R. F. (1984). *Validity and utility of alternative predictors of job performance. Psychological Bulletin, 96,* 72–98.

Ilies, R., Gerhardt, M.W., & Le, H. (2004). Individual differences in leadership emergence: Integrating Meta-analytic findings and behavioral genetics estimates. *International Journal of Selection & Assessment, 12(3),* 207–219.

Jang, K.L., Livesley, W.J., & Vernon, P.A. (1996). Heritability of the Big Five personality dimensions and their facets: A twin study. *Journal of Personality, 64,* 577–591. .

Judge, T.A., Bono, J.E., Ilies, R., & Gerhardt, M. W. (2002). Personality and Leadership: A qualitative and quantitative review. *Journal of Applied Psychology, 87(4),* 765–780.

Judge, T.A., Colbert, A., & Ilies, R. (2004) Intelligence and leadership: A quantitative review and test of theoretical propositions. *Journal of Applied Psychology, 89,* 542–552.

Judge, T.A., Piccolo, R.F., & Ilies, R. (2004). The forgotten ones? The validity of consideration and initiating structure in leadership research. *Journal of Applied Psychology, 89(1),* 36–51.

Jung, C.G. (1971; original in German, 1921). *Psychological types (Collected works of C.G. Jung, volume 6). (3rd ed.).* Princeton, NJ: Princeton University Press.

Kenny, D., & Zaccaro, S. (1983). An estimate of variance due to traits in leadership. *Journal of Applied Psychology, 68,* 678–685.

Lim, K.H., Chan, K., Ng, K.Y., & Lua, L. (2005). *Design components of systematic leadership development.* Paper presented at the 2005 International Military Testing Association conference, Singapore.

Lord, R.G., De Vader, C., & Alliger, G.M. (1986). A meta-analysis of the relation between personality traits and leadership perceptions: An application of validity generalization procedures. *Journal of Applied Psychology, 71,* 402–410.

Lord, R.G., Foti, R.J., & De Vader, C. (1984). A test of leadership categorization theory: Internal structure, information processing, and leadership perceptions. *Organizational Behavior and Human Performance, 34,* 343–378.

Mann, R.D. (1959). A review of the relationship between personality and performance in small groups. *Psychological Bulletin, 66(4),* 241–70.

McCauley, C.D., Moxley, R.S., & Van Velsor, E. (1998). *Handbook of leadership development.* Greensboro, NC: The Center for Creative Leadership.

McCrae, R.R., & Costa, P.T. (1989). Reinterpreting the Myers-Briggs Type Indicator from the perspective of the Five-Factor Model of Personality. *Journal of Personality, 57(1),* 17–40.

McClelland, D.C. (1973). Testing for competence rather than for "intelligence". *American Psychologist, 28,* 1–14.

Montebello, A.R. (2001). Review of "Effective Competency Modeling and Reporting: A step-by-step guide for improving individual and organizational performance". *Personnel Psychology, 54(1),* 1034–1037.

Mumford, M.D., Zaccaro, S.J., Harding, F.D., Jacobs, T.O., & Fleishman, E.A. (2000). Leadership skills for a changing world: Solving complete social problems. *Leadership Quarterly, 11*, 1–18.

Northouse, P.G. (2007). *Leadership Theory and Practice*. Thousand Oaks: Sage Publications.

Popper, M., Amit, K., Gal, R., Mishkal-Sinai, M., & Lisak A. (2007). *The leadership formula: P x M x D* [ARI Contractor Report 2007–10]. Arlington, VA: U.S. Army Research Institute.

Rubin, R.S., Bartels, L.K., & Bommer, W.H. (2002). Are leaders smarter or do they just seem that way? Exploring perceived intellectual competence and leadership emergence. *Social Behavior and Personality: an international journal, 30(2)*, 105–118.

Saucier, G., & Ostendorf, F. (1999). Hierarchical subcomponents of the Big Five personality factors: A cross-language replication. *Journal of Personality and Social Psychology, 76*, 613–627.

Schmidt, F.L., & Hunter, J.E. (1998). The validity and utility of selection methods in personnel psychology: Practical and theoretical implications of 85 years of research findings. *Psychological Bulletin, 124*, 262–274.

Schriesheim, C.A., & Kerr, S. (1974). Psychometric properties of the Ohio State Leadership scales. *Psychological Bulletin, 81*, 756–765

Shartle, C.L., Stogdill, R.M., & Campbell, D.T. (1949). *Studies in Naval Leadership*. Columbus, Ohio: Ohio State University.

Stogdill, R.M. (1948). Personal factors associated with leadership: A survey of the literature. *Journal of Psychology, 25*, 35–71.

Stogdill, R.M., Shartle, C.L. (1948). Methods for determining patterns of leadership behavior in relation to organization structure and objectives. *Journal of Applied Psychology, 32*, 286–91.

Stogdill, R.M., Wherry, R.J., & Jaynes, W.E. (1953). *Patterns of leader behaviour: A factorial study of navy officer performance*. Columbus: Ohio State University.

Terman, L.M. (1904). A preliminary study in the psychology and pedagogy of leadership. *Journal of Genetic Psychology, 11*, 413–451.

Tupes, E.C., & Christal, R.E. (1961). *Recurrent personality factors based on trait ratings*. USAF ASD Tech. Rep. No. 61–97, Lackland Airforce Base, TX: U. S. Air Force.

U.S. Army. (2006). *Army leadership: Competent, confident & agile* (Field Manual 6-22). Washington, D.C.: Headquarters, Department of the Army.

Vernon, P.E. (1979). *Intelligence: Heredity and environment*. San Francisco: W. H. Freeman & Company.

Wiggins, J.S., & Trapnell, P.D. (1997). Personality structure: The return of the big ¢ve. In R. Hogan, J.A. Johnson, & S.R. Briggs (Eds.), *Handbook of personality psychology (pp. 737–758)*. San Diego, CA: Academic.

Yukl, G.A.(1999) An evaluative essay on current conceptions of effective leadership. *European Journal of Work & Organizational Psychology, 8(1)*, 33–48.

Yukl, G.A., Gordon, A., & Taber, T. (2002). A hierarchical taxonomy of leadership behaviour: Integrating a half century of behavior research. *Journal of Leadership and Organizational Studies, 9(1)*, 15–32.

Zaccaro, S.J., Kemp, C.F., & Bader, P. (2004). Leader traits and attributes. In J. Ankonakis, A.T. Cianciolo, & R.J. Sternberg (Eds.), *Nature of Leadership*. Thousand Oaks, CA: Sage.

Zaccaro, S. J., Foti, R. J., & Kenny, D. A. (1991). Self-monitoring and trait-based variance in leadership: An investigation of leader flexibility across multiple group situations. *Journal of Applied Psychology, 76*, 308–315.

Zuckerman, M. (1991). *Psychobiology of personality*. New York: Cambridge University Press.

6

Military Leadership — A Matter of Styles or Values?

Source: CyberPioneer, Ministry of Defence, Singapore

OVERVIEW

This chapter continues to focus on the *leader* in the leadership process and aims to provide an overview of the styles and values approaches to leadership. For some, the idea of styles and values may seem to present opposite views of "how to lead" in the sense that on one hand, the styles approach absent of values may present a somewhat manipulative view of leadership by suggesting that leaders should be flexible in employing different leadership behavioral styles under different situations to achieve certain desired outcomes. In contrast, the values approach would suggest that leaders should be fundamentally and consistently anchored on a set of values and not adapt their actions to manipulate people to achieve whatever outcome that is desired. So, is effective military leadership more a matter of adapting different styles to different situations, or, of consistently upholding a set of principles and values, or both? This chapter begins by summarizing some of the key theories and lessons learned from the styles approach to leadership and how the idea of styles is referenced in various military leadership doctrines today. The next section focuses on the values aspect of military leadership where three approaches are introduced, namely, the character perspective, the ethical reasoning perspective, and the "values-based leadership" perspective.

EARLY "STYLES APPROACH": 1940s–70s

Origins of "Styles" Approach

The scientific study of leadership styles began during the period of the Second World War with a classic experiment by Lewin, Lippitt & White (1939) in which they assigned actors to role-play three types of leaders for groups of school children who were tasked to produce art and craft items for various arts clubs. The three leadership styles were (a) the "authoritarian" leader who remained aloof and gave orders without consulting the group; (b) the "democratic" leader who offered guidance, encouraged and participated in the group's activities; and (c) the "laissez-faire" leader who explained to the children the requirements but did not become involved or participate in the activities. The children in the different experimental groups were matched for intelligence, popularity and "energy".

Lewin et al. (1939) observed that group morale was highest under the democratic leader and that the relationship among the group members and with the leader was friendly. Moreover, when the leader left, the group continued to work independently. On the other hand, they observed two types of behaviors among the children under an authoritarian leader — aggressive children who were rebellious and constantly demanding attention from the leader or children who would blame other members when things went wrong and apathetic children who made less demands and were less critical of the leader. While the "authoritarian" group produced more than the democratic group, the quality of its art and craft items was poor and lacking in originality. Lewin et al. (1939) reported that the groups under the laissez-faire leader performed poorest in both quantity and quality; that the children's morale was weakest and that they cooperated little and demanded most from leader while showing little ability to work independently.

What Are Leadership Styles?

In a well-established textbook of leadership, Northouse (2007) defines leadership styles as "the behavior pattern of a person who attempts to influence others" (p. 93). Unlike the trait approach which emphasizes the leader's personality, the style approach emphasizes descriptive patterns of the leader's behavior that characterize "how different leaders lead". Some consider the styles approach as an extension of the behavioral approach to leadership which was dominated by the "Ohio State" (Shartle, Stogdill & Campbell, 1949) and University of Michigan (Cartwright & Zander, 1960; Katz & Kahn, 1951) studies which established the importance of "Task" and "People" behaviors in leadership in the 1940s and 50s.

Continuum or Grid of Different Styles? During the 1950s and 60s, different models of "task- and people-oriented" leadership styles emerged, with each proposing different ways in which the two styles or orientations could be related. Drawing on the ideas of Lewin et al. (1939), Tannenbaum and Schmidt (1958) in the *Harvard Business Review* proposed a "continuum of styles" that ranged from "boss-centered" or "task-oriented" to "subordinate-centered" or "relationship-oriented". While this "continuum" approach presented Task and People as two opposite styles of leading in the extreme, it also emphasized that leaders could "mix" the degree to which they are task- or people-oriented depending on the situation. Tannenbaum and Schmidt stressed that leaders should consider various situational factors such as the competency and personality of the subordinates, the time pressure of

FIGURE 6-1. Tannenbaum and Schmidt's one-dimensional "continuum" versus Blake and Mouton's two-dimensional "grid" representation of the relationship between "task" and "people" leadership orientations.

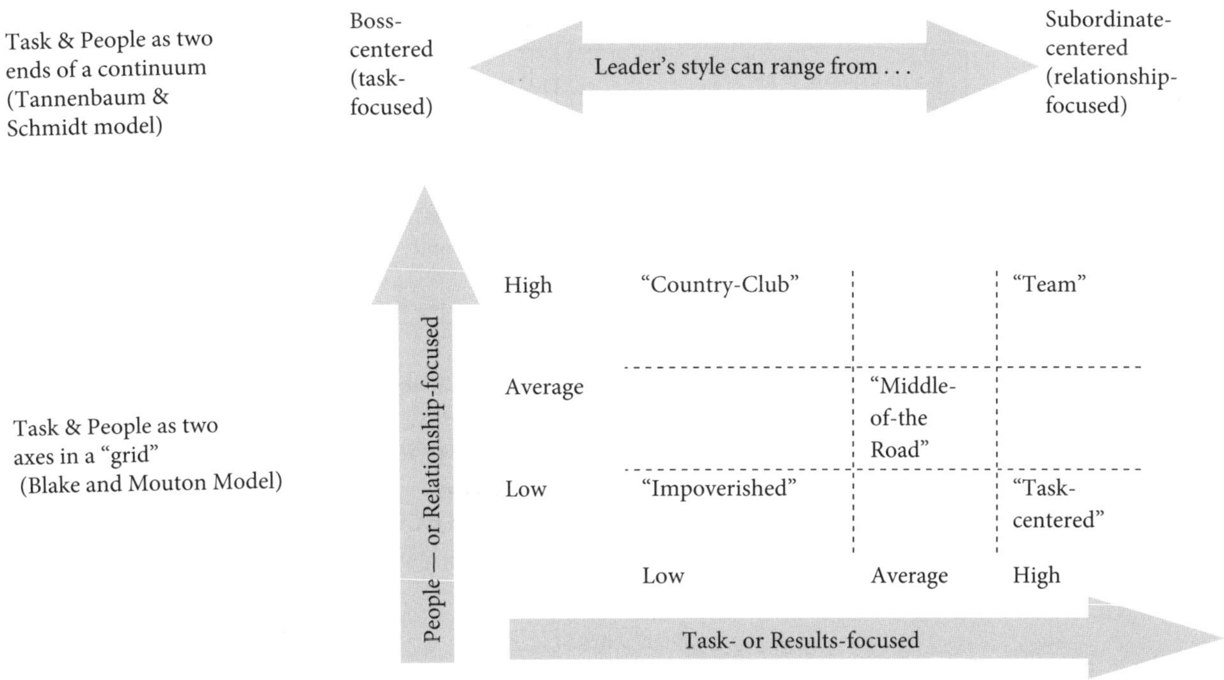

the task, when selecting the most appropriate style for the situation. In 1964, Blake and Mouton also proposed a "Managerial Grid©" theory of leadership styles that borrowed the ideas of "Task" and "People" behaviors. Their idea was that one could characterize leaders into different leadership styles on a "grid" using these two dimensions which they labeled "Concern for People" and "Concern for Results". Examples of the different leadership styles in this grid were the "Team Management" style (high results, high people), "Country-Club" style (high people, low results), "Authority-compliance" style (high results, low people), and "Middle-of-the-road" style (moderate people and results). Figure 6-1 contrasts Tannenbaum & Schmidt's continuum with Blake and Mouton's "Grid" models of the relationship between Task and People orientations.

Situational Leadership. Another popular theory of leadership styles is Hersey & Blanchard's (1969) "Situational Leadership" which proposed a continuum of four styles which varied in task versus people emphases which they labeled "Directing" (high task), "Coaching" (high task and high people), "Supporting" (high people) and "Delegating" (low task or people). They suggested that leaders should change their style along this continuum to meet the "maturity" or developmental level of the subordinates, which they labeled as D1 (Low competence, high commitment subordinates), D2 (somewhat competent but low commitment), D3 (moderate competence but variable commitment), and D4 (high competence and commitment). Figure 6-2 summarizes this model. Situational Leadership

FIGURE 6-2. Hersey & Blanchard's (1969) Situational Leadership Theory suggests "if-then" rules for various styles depending on the followers' "maturity" or "developmental level".

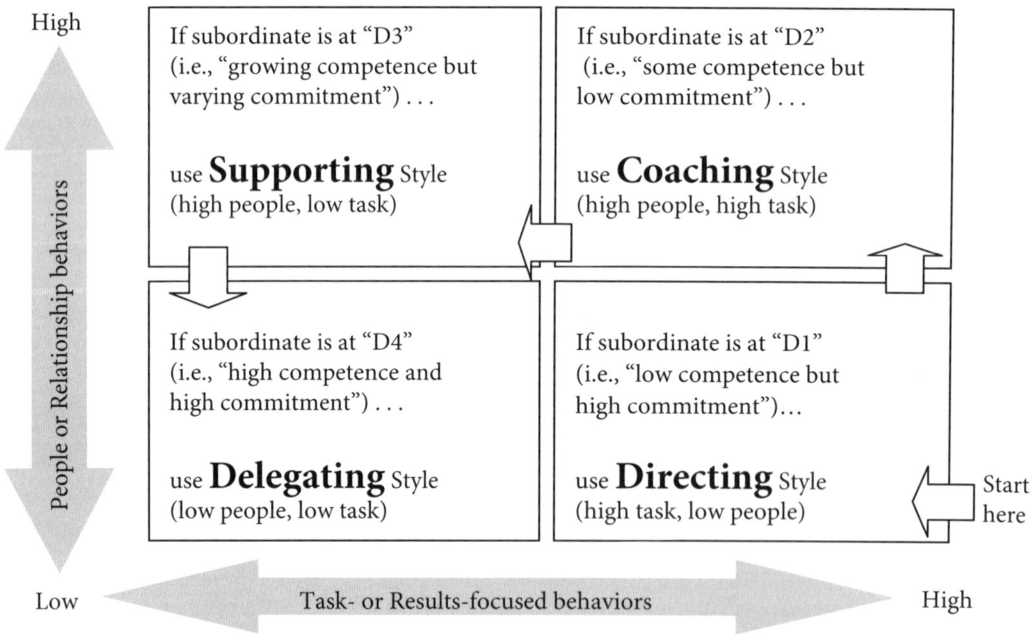

remains a popular model used in leadership training because of its intuitive appeal and its practical nature in that it prescribes specific styles leaders may adopt in different situations. Unfortunately, however, except for a study by Vecchio (1987) that showed that the Directing style was most appropriate for D1 situations, there is a lack of good empirical support for other aspects of the theory. Many have noted the theory is unclear about how subordinate competence and commitment combine to determine their state of "developmental need" or "maturity". It is also unclear why the subordinates should be assumed to have high commitment in D1 and then low commitment in D2. Finally, it is unclear whether leaders should adjust their style to suit entire groups or to each individual with a possibly different developmental level in the group.

Contingency Theory. In contrast to Hersey and Blanchard's Situational Leadership which assumed that leaders could adjust their leadership style according to different situations, Fiedler (1964, 1967) proposed a "Contingency Theory" of leadership. Fiedler believed that a leader's task versus relationship "style", was a function of leader's personality and motivation, and was therefore a fixed "trait" that could not be adjusted to situations. In his view, it was easier for leaders to change their situation to achieve effectiveness rather than change their leadership style. He therefore argued that leadership effectiveness depended on the "matching" of the leader's style with the situational characteristics. He highlighted three factors in any leadership "situation". These were: (1) leader-member relations or the degree to which a leader is accepted and supported by the group members, (2) task structure or the extent to which the task is structured and defined with clear goals and procedures, and (3) position power or the leader's authority to control the subordinates through rewards and punishments. In his view, these three factors together affected overall "situational favorability" where high levels of leader-member relations, task structure and position power was treated as the "most favorable" situation for a leader, while low levels of the three factors were seen as "least favorable". His theory, which was reasonably well supported by empirical studies, was that task-oriented leaders are most effective at either end of the scale while relationship-motivated leaders are most effective in moderately favorable situations. Figure 6-3 summarizes the main ideas in Fielder's contingency theory.

FIGURE 6-3. Main ideas in Fielder's Contingency Theory of Leadership.

<table>
<tr><td colspan="9">Situational factors</td></tr>
<tr><td>Leader-member relations</td><td colspan="4">Good</td><td colspan="4">Poor</td></tr>
<tr><td>Task structure</td><td colspan="2">Structured</td><td colspan="2">Unstructured</td><td colspan="2">Structured</td><td colspan="2">Unstructured</td></tr>
<tr><td>Positional power</td><td>Strong</td><td>Weak</td><td>Strong</td><td>Weak</td><td>Strong</td><td>Weak</td><td>Strong</td><td>Weak</td></tr>
<tr><td>Overall situational "favorability"</td><td colspan="8">"Highest" ⟷ "Lowest"</td></tr>
<tr><td>Leader "style" that would yield best performance</td><td colspan="3">Task-motivated</td><td colspan="3">Relationship-motivated</td><td colspan="2">Task-motivated</td></tr>
</table>

Questions from the Early "Style" Approach. With a proliferation of various situational or contingency theories in the 1960s and 70s, some of which were commercially popular because of their intuitiveness but lacking in empirical support, researchers became somewhat divided in terms of the different assumptions and features of the different theories. For example, should one assume that styles are "fixed" like traits, or that people can flexibly employ different styles across different situations? Was it meaningful to reduce "situations" into single dimensions of "developmental readiness" or "favorability"? What were the ways in which "situations" differed so as to affect the effectiveness of different styles? By the 1980s, interest in the early styles approach diminished with the rise of a "New Paradigm" of leadership that challenged what seemed to be a transactional assumption about the nature of leadership.

THE "NEW PARADIGM" FOR LEADERSHIP STYLES

From Transactional to Transformational Leadership

When behaviorism dominated the social-behavioral sciences from the 1930s to the 70s, most leadership theories or research treated leadership as a set of behaviors that could reinforce or influence groups or followers to achieve desired outcomes. For example, the "Ohio State" and University of Michigan studies that identified the task and people dimensions of leadership in the 1950s treated leadership as an exchange between leaders and followers. In similar fashion, Hersey & Blanchard's (1969) Situational Leadership and Fiedler's (1964, 1967) Contingency theory tried to prescribe styles for different situations using "if-then" rules. For example, if leaders felt that plans, directions and role clarity were lacking in the group, then they should demonstrate more "task behaviors" like planning, directing, or clarifying expectations and roles. On the other hand, if the leaders could identify the specific needs of the followers that would make them satisfied or productive, they would be advised to demonstrate more "consideration" that would raise the follower's motivation. Leadership was thus seen as an "exchange" or a "transaction" of behaviors and rewards or punishments with group members to achieve desired outcomes.

The "New Paradigm" (Bryman, 1992) of transformational and charismatic leadership theories emerged in the 1970s in response to the traditional, transactional or "traited" views of leadership. The term "transformational leadership" was first coined by the sociologist, Downton (1973) in his book entitled *Rebel Leadership: Commitment and Charisma in a Revolutionary Process.* In 1978, James McGregor Burns developed the concepts of transacting and transforming leadership in his book, *Leadership*. Subsequently, the psychologist Bernard Bass (1985) further developed these concepts into his more empirically based transformational-transactional model of leadership.

Burns on "Transforming Leadership"

In 1978, a new paradigm of leadership emerged when the political-historian James McGregor Burns published his book entitled *Leadership*, in which he contrasted "transacting" versus "transforming" leadership. This book was ground-breaking because it refocused the study of leadership from merely describing or prescribing a set of behaviors or exchanges to achieve task goals in a group situation, to introducing the idea that leadership could also involve transforming the underlying motivations, including the needs, values and beliefs of the followers. Breaking from the behaviorist tradition, Burns (1978) did not describe transforming leadership in terms of a set of behaviors. Instead, he suggested that leadership was about "leaders inducing followers to act for certain goals that represent the values and the motivations — the wants and needs, the aspirations and expectations — of both leaders and followers" (Burns, 1978, p.19).

Fundamental to Burn's (1978) idea was that leadership cannot be separated from the idea of a collective purpose, and that effective leadership results in social changes. In this sense, for Burns (1978), transforming leadership ". . . occurs when one or more persons engage [original italics] with others in such a way that leaders and followers raise one another to higher levels of motivation and morality (1978, p. 20). While transacting leaders approach followers with a desire to exchange one thing for another, transforming leaders are those that appeal to the "higher order" ideas or values of their followers and try to get the followers to see the larger social or organizational interests beyond their immediate, personal or short-term interests. "The leader's fundamental act is to induce people to be aware or conscious of what they feel — to feel their true needs so strongly, to define their values so meaningfully, that they can be moved to purposeful action" (p. 43–44; Burns, 1978). For Burns, transforming leadership was therefore concerned with leading through values rather than leading through extrinsic incentives or fear. In his view, the result of transforming leadership is a mutual relationship that converts followers to leaders, and leaders into moral agents. Burns felt that leaders were neither born nor made. Instead, he argued that leadership evolved from a structure of motivation, values, and goals.

Transformational leaders influence through values, ideals, and personal example.
Source: H. Norman Schwarzkopf/Getty Images News/Getty Images

Bass' Model of Transformational Leadership

Intrigued by Burns' ideas, the psychologist Bernard Bass (1985) set out to test the ideas empirically. In a research program that was partially funded by the U.S. Army Research Institute and that involved samples from industrial, military and educational settings (see Bass, 1996, 1998), Bass first asked senior executives to describe the behaviors and effects of transformational leaders that they personally knew. From the pool of behavioral descriptions, a questionnaire was constructed and administered to 176 senior U.S. Army officers who were asked to rate their immediate superiors on 73 behavioral descriptions. Statistical analysis of the survey data revealed that the 73 descriptions could be reduced into five dimensions or "factors", of which two were concerned with transactional leadership (i.e., contingent reward and management-by-exception), and three concerned transformational leadership (i.e., inspirational motivation or charisma, individualized consideration and intellectual stimulation). Following many other rounds of empirical studies in different settings, Bass in 1985 developed the Multifactor Leadership Questionnaire (MLQ) which eventually included over 141 leadership descriptions that were classified by trained judges either as transformational, transactional, or laissez-faire leadership. The MLQ provided scores on four dimensions of transformational leadership and three dimensions of transactional leadership as follows:

- **Transformational leadership** dimensions:
  - *Idealized Influence or Charismatic Leadership.* This included behaviors such as providing followers with a clear vision, sense of purpose, and by role modeling values and ideas.
  - *Inspirational Motivation.* This included behaviors such as motivating others, communicating high expectations, generating enthusiasm and challenging people.
  - *Intellectual Stimulation.* This included behaviors such as soliciting new ideas and new ways of doing things; stimulating others to be creative and never publicly correcting or criticizing others.
  - *Individualized Consideration.* This included behaviors such as establishing a supportive climate where individual differences are respected, where the leader shows a genuine concern for the unique needs and acts to developing others.

- **Transactional leadership** dimensions:
  - *Contingent reward.* This refers to behaviors where the leader assigns work and then rewards the follower for carrying out the assignment.
  - *Management-by-exception.* This includes behaviors where the leader monitors the follower, and then corrects him/her if necessary. Management-by-exception can be either:
    - *Passive.* This style involves waiting passively for errors to occur and then taking corrective action; or,
    - *Active.* This refers to the style of actively supervising and looking out for errors or problems before they arise. Such a style is especially needed when safety is a key concern.

- **Laissez-faire leadership.** Also called "non-leadership", this refers to a "hands-off" and "let things ride" approach where the leader does not attempt to influence the followers or to address their needs.

In 1985, Bass presented some research studies that supported his argument. While transactional leadership could yield performance to requirements, transformational leadership could result in performance "beyond expectations" such that followers would put in more effort and show greater commitment to the task and because they were not just performing for self-interest but for the larger good. While Burns (1978) described transforming and transacting leadership as opposites, Bass (1998) argued for an *additive* relationship between transformational and transactional leadership such that transformational leadership does not substitute for transactional leadership but *augments* transactional behaviors to yield better leadership outcomes. In other words, for Bass (1998), the best leaders are *both* transformational *and* transactional, while the poorest form of leadership is non-leadership or laissez-faire leadership. Describing his model as the "Full Range of Leadership", Bass (1998; see also Bass & Avolio, 1994) argued that every leader can display all of the above "styles" to some extent, and, that an optimal leader would practice the transformational components more frequently and the transactional components less frequently. The important point to note is that transactional leadership is not "bad" or "poor" leadership. Nor should one assume that transformational leadership can replace transactional behaviors. Figure 6-4 illustrates the main ideas in Bass' model of leadership. A

FIGURE 6-4. Main ideas and findings related to Bass' (1995, 1998) theory of leadership.

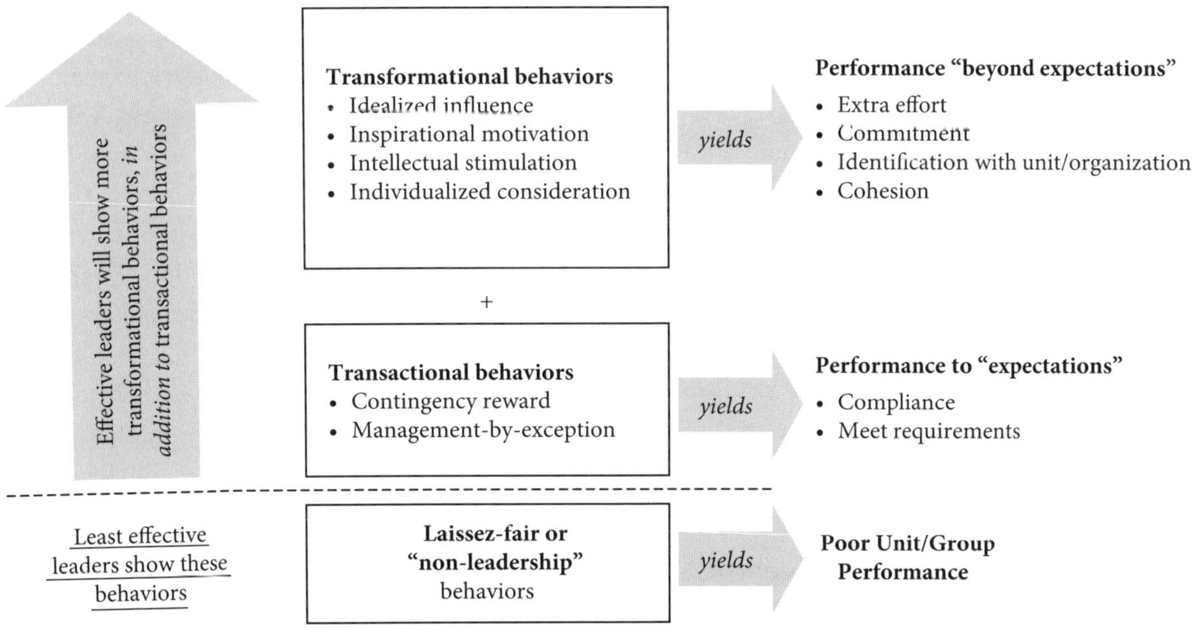

meta-analysis by Judge and Piccolo (2004) generally confirmed Bass's theory and showed that while contingent reward had positive relationships with follower motivation, job satisfaction, satisfaction with their leader, and rated leader effectiveness, transformational leadership had stronger positive effects on these outcomes than transactional leadership behaviors. Their meta-analysis also supported Bass' "augmentation hypothesis".

Military Research Supporting Bass' Theory

In 1992, a study of U.S. Air Force squadron leaders by Curphy showed that transformational and transactional leadership behaviors could positively predict the motivation, cohesion, and performance levels of squadrons. By the mid-1990s, two meta-analyses conducted by Lowe, Kroeck, and Sivasubramaniam (1996) and Patterson, Fuller, Kester, and Stringer (1995) confirmed the positive relationship between transformational leadership and performance reported in the literature. However, these two meta-analyses were criticized in that they were based on studies where the leadership and performance data were either collected at the same point in time or from the same person or both, which could artificially inflate the reported correlations between leadership ratings and performance.

Transactional and Transformational Leadership both Affect Unit Performance. In 2003, Bass, Avolio, Jung and Berson reported a U.S. Army study that showed that the leadership ratings collected from units operating under stable conditions could predict the subsequent performance of those units operating under high stress, challenge and uncertainty. Specifically, transformational and transactional leadership ratings were collected from 72 U.S. Army platoons during normal operational assignments. These ratings were then used to predict the performance of the platoon units that were participating in a high stress, two-week training exercise that was designed to test the unit's effectiveness. The study showed that *both* the transformational and transactional contingent reward leadership ratings of the platoon leaders and sergeants positively predicted unit performance in the exercise, and, that this relationship was partly because these behaviors raised the level of cohesion in the platoons, which in turn increased the effectiveness of the platoons. The study also showed that transactional leadership was important to establish clear standards and expectations of performance among the soldiers, which was in turn important to establish a "base level of trust" in their platoon leader. This base level of trust was in turn reinforced as the soldiers observed their leaders and peers execute their assigned tasks according to standards and expectations. Transformational leadership would then build on these initial, "base-levels" of trust by establishing a deeper sense of identification among followers with respect to the unit's values, mission, and vision.

Full range of leadership can be trained. Avolio and Bass (1994) designed and implemented "Full Range of Leadership" training programs conducted for executives in both profit and non-profit organizations by focusing on the above behaviors. Their follow-up studies after six months to a year showed that the executives improved in their transformational leadership and were able to reduce their use of management-by-exception (MBE). In the late 1990s, the Israeli Defense Force (IDF) School for Leadership Development supported a large scale, empirical research effort to test the key ideas in transformational leadership theory. In a longitudinal, randomized field experiment involving 54 military leaders, their 90 direct followers, and 724 indirect followers, Dvir, Eden, Avolio and Shamir (2002) showed that military leaders who received transformational leadership training had a more positive impact on their direct followers' development, and on indirect followers' performance than did the leaders in "control" group who received an eclectic mix of leadership training such as discussing "here and now" individual and group processes with a psychodynamic focus. In the experiment, the military leaders participated in a three-day workshop that taught them that transformational and transactional leadership were "different lenses" through which they viewed relationships with followers, and, that transformational leadership could be demonstrated through a set of behaviors. A month and a half after the workshops, just before the leaders assumed their first leadership role, the trainers worked with the experimental leaders in three-hour "booster sessions" to reinforce the transformational leadership principles. Dvir et al.'s (2002) IDF experiment showed that transformational leadership could create higher levels of development and performance among followers than transactional leadership. It also showed that transformational leaders tended to have followers who could be continuously developed to higher levels of motivation, morality, and empowerment. To learn more about methods to develop both transformational and transactional leadership skills, readers are encouraged to refer to Sosik &

Research by Bass et al. (2003) showed that both transformational and transactional contingent reward leadership raised cohesion which increased unit effectiveness.

Source: Cris Bouroncle/Getty Images News/Getty Images

Jung's (2010) *Full Range Development* which is based on two decades of careful research on leadership development techniques and interventions applied in both military and non-military contexts.

WHAT WE NOW KNOW ABOUT "STYLES"

After over a half a century of research, it is now established that there are many different styles of leading, and that these can be organized according to various very broad dimensions as follows: task- and people- or relationship-oriented styles, transactional and transformational styles. There is also general acceptance that whether a certain leadership style will be effective depends on the "situation". In other words, no single leadership style is "best" for all situations. Rather than to debate whether styles are "fixed", learned or flexibly employed by leaders, it is now recognized that different styles can yield different effects or outcomes. For example, a directive style generally works best with new or immature followers. Task-oriented styles (i.e., high on initiating structure) generally enhance the group's task performance, while relations-oriented styles (i.e., consideration) generally enhance follower morale and satisfaction with the leader. Research has also shown that if transactional leadership can yield "performance to standards", adding transformational leadership on top of transactional leadership behaviors can yield in "performance beyond expectations" including greater initiative, extra effort, commitment, change, and creativity (see Bass, 1985). Research also shows that transactional leadership tends to result in higher levels of stress among the followers.

Transformational leadership involves more than just charismatic behaviors. It also involves inspirational motivation, intellectual stimulation and individualized consideration. This implies that one does not only need to be "charismatic" to achieve transformational outcomes when leading people. Finally, it is now clearer how situations are complex and cannot be simplified into a few dimensions; that there are countless "situational" factors that may make it impossible to establish any simple rules for prescribing any single "correct" leadership style for each unique situation. Leaders should however be mindful of the many situational factors that affect the effectiveness of any leadership style. These situational factors include:

- The leader's own beliefs about team member participation and his or her confidence in the capabilities of the members.
- The characteristics of the follower, including the extent to which they are independent or dependent-minded, competent or incompetent, committed or uncommitted.
- The team factors including whether the team has the requisite knowledge for the task, whether it shares the organizational values and traditions, and it can work effectively and cohesively.
- The followers' relationship with the leader or the power and authority.
- The nature of the task, including the level of difficulty, risk of failure, and time pressure or urgency of decision or solution.

MILITARY FORCES ON "LEADERSHIP STYLES"

"Styles" in U.S. Army Doctrine

It is interesting to observe how the U.S. Army has evolved its thinking on the topic of "styles" in its doctrine over the past 40 years. In 1973, the U.S. Army Leadership Handbook defined "leadership styles" as "the manner and approach of providing direction, implementing plans, and motivating people". In its 1990 doctrinal manual, the U.S. Army defined leadership styles as "the personal manner and approach of leading (providing purpose, direction, and motivation). It is the way leaders directly interact with their subordinates" (p. 69). The 1973 Handbook described three "basic" styles of military leadership, which were labeled as follows: "Directing", "Participating" and "Delegating".

In 1990, the U.S. Army published a new Field Manual 22-100 on Army leadership which offered this advice to its leaders on applying styles: "Experience has taught you that you should not deal with all people the same. For example, you know it is not effective to deal with a new soldier the same as you would deal with an experienced tank commander or section chief" (p. 69). The 1990 manual also emphasized the need for leaders to employ different styles flexibly to achieve the desired outcomes: "For years, when people talked about leadership styles, they thought about two extremes – an autocratic style and a democratic style. Autocratic leaders used their legitimate authority and the power of their position to get results while democratic leaders used their personality to persuade, and involved subordinates in solving problems and making decisions. Thinking like this fails to consider the possibility of a leader using different styles and being flexible enough to be autocratic at times and democratic at other times, or to combine the two extreme styles at still other times" (p. 69). The manual stressed the importance for leaders to be themselves (or "authentic"), and yet flexible enough to adjust to the people they lead and to the missions assigned. It suggested four factors that leaders should consider when selecting a "correct style" for a situation: the led, the leader, the situation and communications.

In the 1999 edition of its "leadership" Field Manual (22–100), the U.S. Army added "transformational-transactional" to its list of leadership styles. The manual began its discussion on these styles by quoting Napoleon Bonaparte who said, "A man does not have himself killed for a few halfpence a day or for a petty distinction. You must speak to the soul in order to electrify the man", as a way to describe the difference between these styles. Following very closely with ideas in Bass & Avolio's (1994) Full Range of Leadership model, the 1999 manual went on to explain transformational leadership as follows: "As the name suggests, the transformational style 'transforms' subordinates by challenging them to rise above their immediate needs and self-interests. The transformational style is developmental: it emphasizes individual growth (both professional and personal) and organizational enhancement. Key features of the transformational style include empowering and mentally stimulating subordinates: you consider and motivate them first as individuals and then as a group" (p. 3–17).

Interestingly, the latest U.S. Army leadership doctrine published in 2006 (Field Manual 6-22) has dropped any mention of leadership "styles". However, the main ideas of both transactional and transformational leadership are captured in the leadership competency "leading others", which includes both "providing purpose, inspiration and motivation" (i.e., inspirational motivation under transformational leadership) and "enforcing standards" (i.e., transactional leadership). The leadership function called "develop" also incorporates some transformational leadership behaviors while the leadership function called "achieve" largely includes transactional leadership behaviors. It is not known why the U.S. Army dropped explicit discussion on styles or transformational-transactional leadership. What is clear is that its latest doctrine in 2006 has brought greater focus on the behavioral competencies and values required of leaders.

Other Military Forces on Styles

Canadian Forces (CF). In its leadership doctrine, the CF (2005a) defines styles as broad patterns of influence behaviors that are used by leaders which vary in content, tone, intensity. For the CF, styles may be deliberately selected or shaped to achieve a particular effect, or they may be subconscious and habitual forms of the way in which the leader relates to others. The CF presents a "spectrum of influence behaviors" or styles that is roughly ordered by the amount of control employed by the leader, and where the follower's latitude for discretion generally increases from left to right (see Figure 6-5).

FIGURE 6-5. Canadian Force's Spectrum of Leadership Styles.

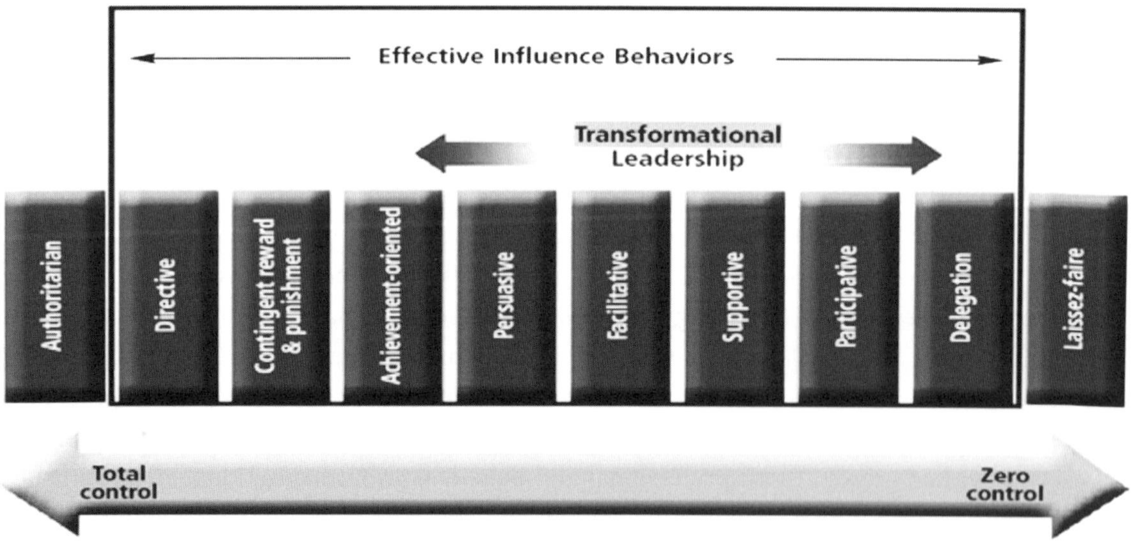

Source: Canadian Forces Leadership Institute, Canadian Defense Academy.

In this spectrum, ineffective styles include the two extremes of "Total Control or Authoritarian" and "Zero Control or Laissez Faire" leadership. The CF (2007) calls for its leaders to demonstrate flexibility in employing the spectrum of styles when leading people: "The most effective leaders are those who can shift between the two styles as required by circumstances and the characteristics of followers" (p. 34).

Australian Defence Forces. In its "Executive pamphlet" on leadership, the Australian Defence Forces (2007) recognizes that existence of different leadership styles and their different impacts on soldiers. "In a practical sense, leaders employ a variety of 'styles' according to the situation within which they find themselves. In an office environment, a supportive style that considers the needs of employees is likely to be more effective than a directive style. Conversely, in a time-critical or ambiguous situation, a directive style that clarifies the path ahead is more likely to motivate confused onlookers than a participative approach" (p. 5–8). More importantly, the Australian Defence Forces (2007) takes a doctrinal position against directive or authoritarian styles, arguing that such styles can prevent their soldiers from developing their moral-ethical development: "When members are encouraged to obey orders for fear of punishment rather than any appreciation of the order's purpose, they are in effect suspending any moral judgment. . . . Autocratic or directive leadership relies on uncritical obedience to external authority and results in dependable task completion — but it also suppresses moral development" (p. 5–9). Doctrinally, the Australian Defence Forces (2007) advocates "a balance in leadership styles rather than an overuse of directive and authoritarian styles", and calls on its military leaders to apply styles that encourage a higher degree of self-discipline or internal rather than external control among its soldiers.

Singapore Armed Forces (SAF). The SAF includes "styles" as one component of its "24-7" leadership framework (see Figure 1-2), and stresses the importance of a flexible employment of a range of styles among its leaders. The SAF Centre of Leadership Development (2010) website explains: "Beyond values and competencies, SAF Leaders should employ a range of leadership styles in order to be effective in a wide range of situations. They must understand how different styles of leading have different impacts and different outcomes" (see also Chan et al. 2005).

VALUES AND MILITARY LEADERSHIP

Military forces have always recognized the importance of values and values inculcation in the military leadership process. Military academies have traditionally placed great emphasis on values inculcation and "character building" as part of leadership training. Contemporary thinking on military leadership continues to emphasize the importance of values and values-based leadership. Most modern military forces today clearly articulate a set of military or "Core" values that they expect all soldiers and military leaders to uphold. The U.S. Army's latest leadership doctrine manual FM6-22 states that "the

Army exists to serve the American people, protect enduring national interests, and fulfill the Nation's military responsibilities. To accomplish this requires *values-based leadership*, impeccable *character*, and professional competence" (italics added, p. 2–3 to 2–4). In the *Conceptual Foundations* for its leadership doctrine, the Canadian Forces (2005b) dedicates an entire chapter to discuss the meaning of values-based leadership. There is also a growing recognition across military forces that discussions on values should not only be concerned about issues of "good versus bad" or "right versus wrong". Instead, military forces increasingly recognize that their leaders and soldiers have to deal with the reality of competing values (e.g., "good versus better", "wrong versus more wrong") that affect their decisions and actions in the complex operating environment.

Why Values Matter

"Life and Death" nature of the Military. One reason for the traditional emphases that military forces place on values lies in the unique nature of its mission and operating context. The Australian Defence Forces (2007) states, "Values such as 'honor', 'duty', 'selfless commitment', 'courage', 'discipline' and 'loyalty' are more prevalent in military organizations than in wider society... These professional military values are concepts derived from the demands of battle. Battlefield situations induce tremendous fear and confusion in individuals. It is no surprise that courage is valued so highly in the military environment since it is courage that is the foil of fear. If someone displays ample courage then the process of influencing them to willingly go into battle is so much easier than if, say, they were imbued with values pertaining to self-preservation" (p. 2–13). The latter point is related to the next reason for military forces to emphasize values — the desire for self-control rather than externally imposed control.

For Internal Control. Another explanation for the military emphasis on values lies in a desire for "internal discipline" or "self-regulation" in contrast to more traditional, externally-imposed rules or "regimental discipline" among soldiers. On its webpage describing its values, the British Army (2008) states: "Being a soldier is not easy. You are asked to do things not asked of other people. You have to be aggressive and strong in battle, yet behave properly and *show self-control* all the time. You have to fit in and be part of a

What kinds of leadership will yield soldiers with internal control and self-regulation?

Source: Scott Olson/Getty Images News/Getty Images

team. You have to trust your teammates and they have to be able to trust you. To do all of this, the Army has six Values that it requires you to live by" (p. 4; italics added for emphasis). Similarly, the Canadian Forces (2005b) in its *Conceptual Foundations* document states that values "tell CF members and their leaders what organizational outcomes they should strive to achieve and what professional standards they should use to guide and regulate their conduct", that "a key proposition of values-based leadership is that the guidance provided by core CF values is the ultimate recourse and compass when explicit direction from superiors is lacking, when dealing with ambiguous situations, or when operating under competing demands and pressures" (p. 16). In its *Executive Pamphlet on Leadership*, the Australian Defence Forces (2007) contrasts values with rules as a method for controlling individual actions in military groups: "The advantage of values over rules as a guide to group behavior is the adaptability they provide in ambiguous situations" (p. 1-11). In other words, for many modern military forces, a clear set of military values provides the basis for internal discipline in soldiers which enables the consistency, initiative, and adaptability needed in its soldiers in the modern military operating environment.

Perspectives on Values and Military Leadership

Character Perspective or "What Leaders Must Be". One way in which military forces articulate the link between values and military leadership is in terms of how values form the foundation of military identity or "character". The U.S. Army has long-emphasized this in the "Be" aspect of its Be-Know-Do leadership framework which also corresponds with its vision for leaders: "The Army Leader: Person of *Character*, Presence and Intellect" (italics added). The U.S. Army (2006) states: "Character, a person's moral and ethical qualities, helps determine what is right and gives a leader motivation to do what is appropriate, regardless of the circumstances or the consequences. . . . Since Army leaders seek to do what is right and inspire others to do the same, they must embody these values" (p. 4–1). Similarly, Chan, Singh, Ramaya & Lim (2005) noted that when the Singapore Armed Forces first promulgated its "Core Values" in 1996, the idea was for the SAF Core Values to be synonymous with the SAF's Character. The first edition of the SAF Core Values handbook stated: "When one thinks of the SAF and an SAF soldier, the qualities associated with its members should exemplify prominently the SAF Core Values" (p. i). The SAF Core Values were also intended to unify members of the SAF: "These values will ensure that there are certain character strengths which bind our people together in fulfilling our professional roles and duties, guiding us toward a common goal" (p. 2).

Ethical Reasoning Perspective or "What One Ought to Do". Besides the character-based perspective of values, the U.S. Army (2006) also states that values are important for leader decision making and choices. "The Army Values consist of the principles, standards, and qualities considered essential for successful Army leaders. They are fundamental to helping Soldiers and Army civilians make the right decision in any situation". It explains the relationship between values and ethics as follows: "Ethics are concerned with how a person should behave. Values represent the beliefs that a person has. . . . the translation from desirable ethics to internal values to actual behavior involves choices" (p. 4–14). The Australian Defence Forces (2007) also places a heavy emphasis on military ethics in its leadership doctrine. "Military ethics has been described as the application of ethics (knowing what one ought to do) to military endeavors. . . . Military leaders will occasionally be confronted by ethical dilemmas on operations and in their careers. To fire upon children brandishing enemy weapons or to interrogate prisoners withholding imminent ambush information are two contemporary examples. By their very nature, dilemmas have no easy answers. Nevertheless, leaders are expected to apply moral sensitivity and be able to ethically justify their decisions and actions" (p. 2–14).

"Values-based leadership". Today, it is also common to find military forces using the phrase "values-based leadership" as part of the continuing effort by their leaders to strengthen the forces' professional and organizational culture (e.g., norms and standards of conduct and behavior or "professional ethic"), climate (e.g., expectations) and identity. The Canadian Forces (CF) expects its leaders to role model its values as part of reinforcing and embedding the values in its culture: "Intentionally or otherwise, leader behavior sends signals about what is centrally important and what is not so important. To the degree that leaders consciously serve, exemplify, and uphold CF values, they affirm through their use the validity and credibility of espoused values, and provide CF members with general guidance as to how they should orient and regulate their own behavior. This is what values-based leadership means . . ." (p. 23, 2005b).

Similarly, Australian Defence Force introduces values-based leadership as a "general leadership notion for any organization where the behavior of leaders reflects their values and sets the example for others in the organization". It stresses how practicing values-based leadership is separate from the content of the values themselves — "Values-based leadership as a concept can apply equally to a terrorist organization or to a hospital. The difference in how outcomes are achieved within these two groups, however, comes down to the values that form the basis of their leadership. Both groups could well share the values of courage, teamwork and initiative. It is unlikely, however, that a terrorist organization would embrace the values of compassion, freedom, tolerance and the respect and dignity of all persons" (p. 1–11).

BEYOND STYLES AND VALUES: "AUTHENTIC LEADERSHIP"

When Burns (1978) first wrote about transforming leadership, he argued that "transforming leadership ultimately becomes moral in that it raises the level of human conduct and ethical aspiration of both leader and the led, and thus it has a transforming effect on both" (p. 20). In contrast, Bass (1985) who treated transformational leadership as a pattern of behaviors, argued that transformational

Authenticity — a vital link between styles and values approaches to leadership.
Source: Stephen Morton/Getty Images News/Getty Images

leaders could be ethical or unethical depending on the nature of their values and motivations. Bass and Steidlmeier (1999) distinguished "authentic" and "pseudo-transformational" leaders. Authentic transformational leaders transform the followers by building a shared sense of values to guide their collective actions. In contrast, pseudo-transformational leaders are those who coerce and manipulate their followers so as to further their self-interest. They considered Hitler a pseudo-transformational leader or "a case study in transformational leadership gone wrong" because although he appealed to the values and ethics of the German people, he was driven more by personal ambition rather than by the collective ideals of the German people.

Today, one of the latest approaches to leadership is the concept of authentic leadership, where "authentic leaders" are described as those who "know who they are, what they believe and value, and . . . act upon those values and beliefs while transparently interacting with others" (p. 803, Avolio, Gardner, Walumbwa, Luthans & May, 2004). Such leaders "are deeply aware of how they think and behave and are perceived by others as being aware of their own and others' values/ moral perspectives, knowledge, and strengths; aware of the context in which they operate; and who are confident, hopeful, optimistic, resilient, and of high moral character" (p. 4, Avolio, Luthans & Walumbwa, 2004).

To date, we are not aware of any military force using the concept of authentic leadership in its doctrine. The U.S. Army (2006), however, emphasizes the importance of authenticity in the process of leadership influence: "To succeed and create true commitment, influencing techniques should be perceived as authentic and sincere. Positive influence comes from leaders who do what is right for the Army, the mission, the team, and each individual Soldier. Negative influence — real and perceived —

emanates from leaders who primarily focus on personal gain and lack self-awareness. Even honorable intentions, if wrongly perceived by followers as self-serving, will yield mere compliance. False perception may trigger unintended side effects such as resentment of the leader and the deterioration of unit cohesion" (p. 7–5).

Concluding Remarks

Having read this Chapter, do you now believe that military leadership is more a matter of styles or values, or both? We hope that readers will recognize from both Chapters 5 and 6 that leadership is in fact a multifaceted phenomenon; that leadership is not only about styles and values, but it also concerns matters of traits and behaviors. Moreover, these four aspects only concern the *leader* in the *leadership* process — the latter involves more than the leader, it also concerns the context of the task or mission, the task or mission, the operating environment, and the outcomes that matter. These ideas are captured in the Singapore Armed Forces' Leadership 24-7 framework which distinguish four aspects of leader development in its "triangle" from what it calls the context of leadership, i.e., the "circle" in the framework (see Chan et al., 2005). The remaining chapters attempt to help readers understand the nature of military leadership in the context of teams in high-stress, high risk and high-tempo military operations, and also in the context of military organizational structures (e.g., hierarchy) and culture (e.g., the "profession of arms").

SUMMARY

The style approach to leadership attempts to describe certain patterns of the leader's behavior that each characterizes "how different leaders lead". Different leadership styles have been proposed and studied. Early research by Lewin et al. (1939) studied Authoritarian, Democratic and Laissez-faire styles of leadership. Various theories in the 1960s and 1970s proposed how certain leadership styles were largely based on the Task versus People/Relations dimensions of leadership behaviors. These included Blake & Mouton's (1964) Managerial Grid, Hersey & Blanchard's (1972) Situational Leadership, and Fiedler's (1967) Contingency theory of leadership. The latter two emphasized how the effectiveness of different styles depended on situational factors. From the 1980s, a new approach to leadership styles emerged with researchers focusing on transformational and charismatic styles of leadership, in contrast with the earlier transactional styles. One of the most dominant models in this new approach is Bass' (1998) "Full Range of Leadership". While many different labels and terms have been proposed to describe different styles of leading, these can generally be organized according to two very broad dimensions as follows: Task versus People/Relations, and Transactional versus Transformational styles. No single leadership style is "best" for all situations. Whether a certain style will be effective depends on the "situation". However, "situations" are extremely complex with countless variables (e.g., leader, follower, task and organizational characteristics), that make it almost impossible to establish any simple rule to prescribe any correct styles any situation.

Today, the U.S. Army no longer discusses styles explicitly in its doctrine. This does not reflect a rejection of the idea of styles, but rather, that the U.S. Army has infused the basic lessons of the styles approach (e.g., importance of leaders flexibly employing a full range of leadership behaviors across different situations to achieve different desired outcomes) into its more competency- and character-focused doctrine of military leadership. Styles continue to feature in the Canadian and Singapore forces' military leadership doctrines. Modern military forces continue to emphasize the importance of values and values-based leadership in their doctrines. They do so for various reasons including the professional or warrior "ethos" that soldiers need to do their jobs and the need for internal rather than external control (e.g., self-discipline versus rule based, "regimental" discipline) among soldiers. There is also a growing recognition that leaders and soldiers increasingly have to adopt a more sophisticated approach to dealing with values and ethics in the complex operating environment in that values are often not just about being "good or bad" or "right versus wrong" issues. Rather, values often matter in dilemmas or decisions where values compete, e.g., "good versus better", "wrong versus more wrong". Military forces also seem to emphasize the importance of values and leadership in three different ways. Some, such as the U.S. Army, emphasize values as part of the desired character of its leaders. Others such as the Australian and Canadian forces emphasize the "ethical approach" which calls for their military leaders to constantly apply ethical reasoning in their actions. Today, many forces also use the phrase "values-based leadership" as part of the constant effort to build and strengthen the military's professional and organizational culture (including norms and standards of conduct and behavior or "professional ethic"), climate (expectations) and identity.

Key Concepts

- Leadership styles
- Situational leadership
- Contingency theories
- Full Range of Leadership
- Values and Ethics
- Character
- Authentic leadership
- Pseudo-transformational leadership
- Ethical/ethical reasoning perspective
- Transformational
  - Individualized consideration
  - Idealized influence
  - Inspirational motivation
  - Intellectual stimulation
- Transactional
  - Contingent Reward
  - Management by exception – active
  - Management by exception – passive
- Laissez-faire

Questions

- When discussing leadership, people commonly ask questions like "Is there a leadership style that is best for me?", or "what leadership style works best for this context?" Having read this chapter, what advice would you give them about how to think about "leadership styles"? Are those common questions appropriate? Why?
- How does your military deal with the issues of "styles" and "values" in the leadership process? Does it treat these as different or related aspects of leadership? Are there prescriptions of any preferred "style" or set of "values"?
- In your view, is "authenticity" an important concept in military leadership? Why? Would you regard it as a behavioral style or something than concerns values, or both? What does it take to be an "authentic" military leader?

References

Australian Defence Headquarters (2007). *Leadership in the Australian Defence Force* (ADDP 00.6). Director Centre for Defence Leadership Studies, Defense Publishing Service, Canberra, ACT.

Avolio, B.J. (1999). *Full Leadership Development.* Sage Publications: California.

Avolio, B.J., & Bass, B.M. (1994). *Evaluate the impact of transformational leadership training at individual, group, organizational and community levels.* Final report to the W.K. Kellogg Foundation, Binghamton University, Binghamton, New York.

Avolio, B.J., Gardner, W.L., Walumbwa, F.O., Luthans, F., & May, D.R. (2004). Unlocking the mask: A look at the process by which authentic leaders impact follower attitudes and behaviour. *The Leadership Quarterly, 15*, 801–823.

Avolio, B.J., Luthans, F., & Walumbwa, F.O. (2004). *Authentic leadership: Theory-building for veritable sustained performance.* Unpublished Monograph. Gallup Leadership Institute, University of Nebraska, Lincoln.

Bass, B.M. (1985). *Leadership and performance beyond expectations.* New York: Free Press.

Bass, B.M. (1995). Theory of transformational leadership redux. *The Leadership Quarterly, 6(4)*, 463–78.

Bass, B.M. (1996). *New paradigm of leadership: An inquiry into transformational leadership.* Alexandria, VA: U.S. Army Research Institute for the Behavioral and Social Sciences.

Bass, B.M. (1998). Leading in the Army After Next. *Military Review, 78(2)*, 46–57.

Bass, B.M. (1998). *Transformational leadership: Industrial, military and educational impact.* Mahwah, NJ: Lawrence Erlbaum Associates.

Bass, B.M., & Avolio, B.J. (1992). Developing Transformation Leadership: 1992 and beyond. *Journal of European Industrial Training, 14*, 23.

Bass, B.M., & Avolio, B.J. (1994). *Improving organizational effectiveness through transformational leadership.* Thousand Oaks, CA: Sage.

Bass, B.M., Avolio, B.J., Jung, D.I., & Berson, Y. (2003). Predicting unit performance by assessing transformational and transactional leadership. *Journal of Applied Psychology, 88(2)*, 207–218.

Bass, B.M., & Steidlmeier, P. (1999). Ethics, character, and authentic transformational leadership behavior. *The Leadership Quarterly, 10(2)*, 181–217.

Bennis, W., & Nanus, B. (1985). *Leaders: The strategies for taking charge*. New York: Harper & Row.

Blake, R.R., & Mouton, J.S. (1964). *The Managerial Grid*. Houston, TX: Gulf Publishing.

British Army (2008). *A British Soldier's Values and Standards* (AC 63812). Retrieved on 1 June 2010 from: http://www.army.mod.uk/documents/general/v_s_soldiers_ed.pdf.

Bryman, A. (1992). *Charisma and leadership in organizations*. London: Sage.

Burns, J.M. (1978). *Leadership*. New York: Harper & Row

Canadian National Defense (2005a). *Leadership in the Canadian Forces*. Kingston, Ontario: Canadian Forces Leadership Institute, Canadian Defense Academy.

Canadian National Defense (2005b). *Leadership in the Canadian Forces: Conceptual Foundations*. Kingston, Ontario: Canadian Forces Leadership Institute, Canadian Defense Academy.

Canadian National Defense (2007). *Leadership in the Canadian Forces: Leading People*. Canadian Forces Leadership Institute, Canadian Defense Academy.

Cartwright, D., & Zander, A. (1960). *Group dynamics*. Evanston, IL: Row, Peterson.

Chan, K., Singh, S., Ramaya, R., & Lim, K.H. (2005). *Spirit & System: Leadership Development for a Third Generation SAF [4th Pointer Monograph]*. Singapore: SAFTI Military Institute.

Conger, J.A. & Kanungo, R.N. (1987). Toward a behavioral theory of charismatic leadership in organizational settings. *Academy of Management Review, 12*, 637–647.

Curphy, G.J. (1992). An empirical investigation of the effects of transformational and transactional leadership on organizational climate, attrition and performance. In K.E. Clark, M.B. Clark and D.R. Campbell (Eds.), *Impact of Leadership*. Greensboro, NC: The Center for Creative Leadership.

DeYoung, C.G., Quilty, L.C., & Peterson, J.B. (2007). Between facets and domains: 10 Aspects of the Big Five. *Journal of Personality and Social Psychology, 93*, 880–896.

Downton, J.V. (1973). *Rebel leadership: Commitment and charisma in the revolutionary process*. Riverside, NJ: Macmillan.

Dvir, T., Eden, D., Avolio, B.J., & Shamir, B. (2002). Impact of transformational leadership on follower development and performance: A field experiment. *Academy of Management Journal, 45*, 735–744.

Gal, R. (1985). Commitment and obedience in the military: An Israeli case study. *Armed Forces & Society, 11*, 553–564.

Fiedler, F.E. (1964). A contingency model of leadership effectiveness. In L. Berkowitz (Ed.), *Advances in experimental social psychology*. NY: Academic Press.

Fiedler, F.E. (1967). *A theory of leadership effectiveness*. New York: McGraw-Hill Book Company.

Hersey, P., & Blanchard, K.H. (1969). *Management of organization behavior*. Englewood Cliffs, NJ: Prentice-Hall.

Howell, J.M., & Avolio, B.J. (1993). Transformational leadership, transactional leadership, locus of control, and support for innovation: Key predictors of consolidated business unit performance. *Journal of Applied Psychology, 78*, 891–902.

Judge, T.A., & Piccolo, R.F. (2004). Transformational and transactional leadership: A meta-analytic test of their relative validity. *Journal of Applied Psychology, 9(5)*, 755–68.

Katz, D., & Kahn, R.L. (1951). Human organization and worker motivation. In L.R. Tripp (Ed.), *Industrial productivity* (pp. 146–171), Madison, WI: Industrial Relations Research Association.

Lewin, K., Lippitt, R., & White, R.K. (1939). Patterns of aggressive behavior in experimentally created social climates. *Journal of Social Psychology, 10*, 271–279.

Lowe, K.B., Kroeck, K.G., & Sivasubramanian, N. (1996). Effectiveness correlates of transformational and transactional leadership: A meta-analytic review. *Leadership Quarterly, 7*, 385–425.

Northouse, P.G. (2007). *Leadership: Theory & practice (4th ed.).* Thousand Oaks, CA: Sage.

Patterson, C., Fuller, J.B., Kester, K., Stringer, D.Y. (1995, April). *A meta-analytic examination of leadership style and selected compliance outcomes.* Paper presented at the 10th Annual Conference of the Society for Industrial and Organizational Psychology, Orlando, FL.

Shamir, B., Zakay, E., Breinin, E., & Popper, M. (1998). Correlates of charismatic leader behavior in military units: Subordinates' attitudes, unit characteristics, and superiors' appraisals of leader performance. *Academy of Management Journal, 41(4)*, 387–409.

Shartle, C.L., Stogdill, R.M., & Campbell, D.T. (1949). *Studies in Naval Leadership*. Columbus, Ohio: Ohio State University.

Shils, E.A., & Janowitz, M. (1948). Cohesion and disintegration in the Wehrmacht in World War II. *Public Opinion Quarterly, 12*, 280–315.

Singapore Armed Forces (SAF) Centre of Leadership Development (2010). *The SAF Leadership Framework: SAF Leadership 24–7.* Retrieved on 16 June 2010 from: http://www.mindef.gov.sg/imindef/mindef_websites/atozlistings/saftimi/units/cld/keyideas/framework.html.

Solomon, Z., Mikulinicer, M., & Hobfill, S.E. 1986. Effects of social support and battle intensity on loneliness and breakdown during combat. *Journal of Personality & Social Psychology, 51*, 1269–1276.

Sosik, J.J., & Jung, D.I. (2010). *Full range leadership development: Pathways for people, profit and planet.* New York: Routledge/Psychology Press.

Vecchio, R.P. (1987). Situational leadership theory: An examination of a prescriptive theory. *Journal of Applied Psychology, 72(3)*, 444–451.

Tannenbaum, R. & Schmidt, W.H. (1958). How to choose a leadership pattern. *Harvard Business Review, 36*, March-April, 95–101.

U.K. Defence Leadership Centre (2004). *Leadership in Defense*. Defense Academy of the United Kingdom, Shrivenham, Swindon.

U.S. Army (1973). *Military Leadership.* Headquarters, Department of the Army, Washington, D.C.

U.S. Army (2006). *Army leadership: Competent, confident, and agile* (FM 6–22). Headquarters, Department of the Army, Washington, D.C.

7

Science and Practice of Leading Military Teams

Source: CyberPioneer, Ministry of Defence, Singapore

OVERVIEW

Chapters 2, 3 and 4 discussed the broad military context and the challenges of leadership, covering several ideas from the field of military psychology. Chapter 5 focused on the leader and discussed "person" factors while Chapter 6 discussed leadership styles (and behaviors) and how behaviors are connected to leadership values. The concepts in the latter two chapters were discussed without making specific reference to the context of leadership. This chapter focuses on leadership in the specific context of "teams". The first section of the chapter begins with a brief discussion on the evolution of team research and explains basic ideas related to "teams", "teamwork" and "team effectiveness". The section also focuses on the science behind the practice of leading teams and presents military-relevant team effectiveness models that include "leadership" as a factor. Leadership processes in a team are explained with the functional leadership perspective and the concept of "team leadership". In addition, the section also highlights some research related to leader behaviors in teams and discusses the importance of building trust in teams and trust in leadership. The second section of the chapter focuses on the practice of leading teams and introduces a specific team intervention referred to as "team building" from the Organizational Development (OD) literature. The section includes a brief history related to team building efforts in the military and the status of research on team building. Team building in the military can be systematically implemented. A team building model and framework for systematic team building are presented. Team leaders will also learn about a related team intervention known as "team development" and be introduced to the different roles that they can take at the various developmental stages of a team.

THE SCIENCE OF LEADING TEAMS

Evolution of Team Research

Military forces such as the U.S. Army and U.S. Navy have always taken teamwork and team training seriously and have been the primary sponsors of team research in the past. The research studies on teams have brought together the work of academics and practitioners and generated a vast amount of knowledge in the area of teams and teamwork. There is more knowledge about teams today as a direct result of the research sponsored by the military in the last two decades. Research on teams evolved from the study of groups and group dynamics. The earliest of such military-sponsored research on groups are the *American Soldier* studies by Stouffer and colleagues (1949) after the Second World War. These studies that focused on soldier performance and cohesion in military units had a significant impact on the military in terms of policy changes in the areas of selection, recruitment, training and deployment of personnel in the military. Some of these ideas are discussed in Chapter 4 (see "Cohesion", "Espirit de Corp" and "Morale".) The basic lessons were that soldiers who have trained and fought together came to share a common bond, were highly cohesive and should remain within their small groups when they are deployed in future operations. The studies showed that highly cohesive teams were better prepared for their missions than less cohesive military teams. Such studies on military cohesion were more relevant to past military operations and large scale wars where soldiers had to work together in large numbers. Today, when most military organizations have transformed, downsized, and reorganized into smaller fighting teams, the research on small teams and team effectiveness have become more important to the military.

Group and team research in the military previously adopted the social-psychological perspective which was the dominant approach in the 1950s to the 1970s. Concepts such as "groupthink" and "group dynamics" were commonly used in the social-psychological literature on groups and teams, and models of group performance were based on social-psychological theories. The basic idea was that a group (and the context of a "group") affected the behavior of individuals differently. Since WWII, other research efforts focused on "work teams" specifically to improve productivity and reduce human errors in the workplace. Military forces today are highly dependent on technology and have introduced sophisticated weapons that are guided by computer-information systems. With the increased use of technology and the potential for human errors, the study of teams and teamwork in the military took on a multi-disciplinary perspective involving the field of ergonomics and human factors as well as cognitive psychology. Human factors research began to include studies on how well teams communicated and made decisions to minimize errors.

In the 1970s to 1980s, the aviation industry suffered the consequences of serious airline disasters. Analyses of these accidents showed poor crew communications as a key factor. Similarly, the U.S. military in the 1980s sponsored research in the area of team performance when it experienced a series of accidents involving naval warfare teams and tank crews (Salas, Bowers & Cannon-Bowers, 1995). The research and training that resulted from Crew Resource Management (CRM) studies brought attention to not only the behavioral but also the cognitive and affective aspects of team performance in a technologically complex environment. CRM training was subsequently implemented by the U.S. Navy in 1986 (Salas, Wilson, Burke, Wightman & Howse, 2006).

Team research today brings together subject-matter experts from different domains — researchers, practitioners and even engineers of the technological systems as was witnessed in the Tactical Decision Making Under Stress or TADMUS program, a research program which was initiated after the USS Vincennes incident. In July 1989, the USS Vincennes, a guided missile cruiser, accidently shot down an Iranian Airliner and killed 290 people (Cannon-Bowers & Salas, 1998). The program sponsored by the Office of Naval Research (ONR) was a major effort to understand how teams made decisions and performed under extremely stressful and demanding conditions. Scientists and researchers in the TADMUS research program applied knowledge from the available academic literature on team effectiveness. The TADMUS studies continue to provide valuable information on how to better train and prepare military teams for high-stress, high-demand

Military teams are by nature interdependent.
Source: Rubberball/Rubberball Productions/Getty Images

situations that military teams will increasingly face in the future. The TADMUS researchers have also developed and tested team training approaches that have become important starting points for team research and application in the military. These research studies will be discussed in the next chapter.

Basic Terminology

Teams. Salas, Dickinson, Converse, & Tannenbaum (1992) defined a team as "a distinguishable set of two or more people who interact, dynamically, interdependently, and adaptively toward a common and valued goal, objective or mission, who have been assigned specific roles or functions to perform, and who have a limited life-span of membership". The features of "interdependence", "common and valued goal, objective or mission" and "adaptability" are important and relevant especially to the military context (Salas, Cannon-Bowers, Church-Payne & Smith-Jentsch, 1998). Military teams are, by nature, interdependent — i.e. even though team members have roles that are highly specialized, members are dependent on each other's inputs which need to be coordinated toward a team task or a critical team decision. A "common goal or mission" is the very reason why teams exist in the military. "Adaptability" (and "adaptive expertise in teams" which will be discussed in the next chapter) has been highlighted as an important competency for 21st century military teams.

Teams vs. Groups. In their textbook on Leadership, Hughes, Ginnet & Curphy (2002, p.300) distinguished a "team" from a "group" by highlighting the following features:

- A team has a stronger sense of identification amongst its members while a group's identity may be less clear.
- A team has a common objective or goal while members of a group might not have consensus about the goal.
- Task interdependence is higher in teams while group members can achieve goals by working independently.
- Team members have roles that are more differentiated and specialized than group members.

Military teams often develop a strong sense of identity and military organizations consciously build and strengthen this identity at the primary group-level. Identity at the primary group is an important factor in the development of cohesion and esprit de corps at the larger organizational level. Military teams are given specific missions that provide direction to their day-to-day tasks as well as shape their performance and training goals. Military tasks are by far too complex to be carried out by individuals alone and therefore the need for a high degree of interdependence.

Teamwork. The term "teamwork" has commonly been used to refer to team interactions in a work group setting. A simple approach to "teamwork" has been to focus on just interpersonal relationship issues in a team or to improve communication between team members in order to enhance team performance. Team research from the last two decades revealed that there is more to teamwork than just team member interactions or communication between members. Salas, Rosen, Burke & Goodwin (2009) defined "teamwork" as "the means by which individual task expertise is translated, magnified and synergistically combined to yield superior performance". Salas et al. (2009) suggest that "team leadership", "adaptability", "mutual performance monitoring", "backup behavior" and "team orientation" are the "big five factors" in teamwork and that the coordinating mechanisms responsible for these factors are "shared mental models", "closed-loop communication" and "mutual trust". The research on teamwork is far more complex and not so commonsensical as will be explained later in this chapter. The topic of shared mental models will be covered in depth in the next chapter while the importance of building trust in teams will be discussed in this chapter.

Models of Team Effectiveness

Input-Process-Output (I-P-O) Approach. The discussion of team effectiveness often begins with a description of the Input-Process-Outcome or I-P-O approach by McGrath (1964). Figure 7-1 is a representation of the I-P-O model. McGrath identified individual variables (e.g., personality and attitudes to working in the team), the group factors (e.g., structure, size and cohesiveness) and the environmental factors (e.g., the task characteristics, the reward structure and stress) as comprising the input stage which influences group interactions in the process stage. The process stage is described as a stage where group members manage and overcome interpersonal conflicts to focus on the task or goals of the group. Group interactions determine the outcomes of the team. The team achievements

FIGURE 7-1: A representation of the I-P-O model of team effectiveness.

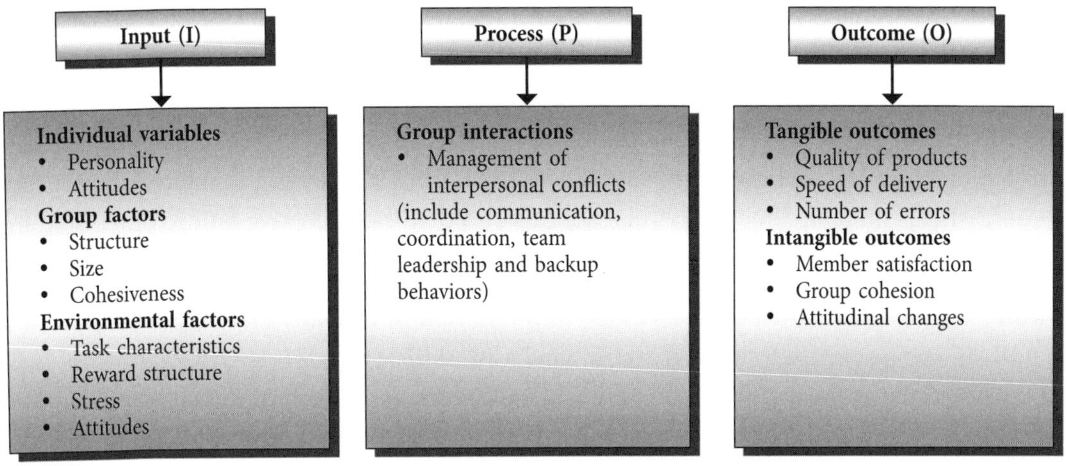

in the outcome stage include both tangible and intangible outputs of the team. The tangible outputs are performance-related variables that can be measured in terms of "quality of products", "speed of delivery" and the "number of errors" while intangible outcomes include "member satisfaction", "group cohesion" and "attitudinal changes". The intangible outcomes would then determine how the same team would perform in the future. The I-P-O approach to understanding team effectiveness suggests that causal relationships exist between outcomes, inputs and process variables. It is the understanding of these causal relationships that provides clues to what makes teams effective.

Teamwork vs Task work Processes. In 1984, Foushee, one of the first researchers to adopt the I-P-O approach to understanding teams, examined teamwork in the cockpit and identified process variables such as adaptability, communication, coordination, decision-making, team leadership, situation awareness, workload management, and backup behaviors. His study led to subsequent models differentiating task work and teamwork and depicted relationships between input, process and outcome variables. Task work refers to the technical aspects of work that need to be carried out by the team. Teamwork refers to the interaction of team members or how well the team carries out its team processes like communication, coordination and supportive actions. These are also known as "process variables" which determine the relationship between input and output variables in the I-P-O model. Foushee's 1984 study resulted in a better understanding and appreciation of the dynamic and multidimensional nature of teamwork. The list of teamwork or process variables relevant to the military context included "adaptibility", "mission analysis and planning" and "assertiveness" (Prince and Salas, 1993).

Process Gains vs Process Loss. Another approach to understanding team processes is the "value-added" by the team's interactions that can result in either process gains or process loss. In 1987, Driskell, Salas & Hogan proposed a model of team effectiveness in which they argued that the added value of the team lies in the team processes. For example, communication structures and task performance strategies, and that the interaction of input factors and team processes can result either in process loss (e.g., social loafing) or process gain (pooled efforts). Process losses are due to team dysfunctions that lead to team ineffectiveness while process gains result from positive team member interactions that lead to team effectiveness. Similarly, Zaccaro, Rittman & Marks (2001) highlighted that the "challenge of integration" was one of the three fundamentals of effective team performance (the other two being "complex environments" and "team leadership"). They highlighted that team members need to successfully integrate their individual actions and that the collective failure of team members to "coordinate and synchronize" their individual contributions may lead to team failure (i.e., process loss).

Team Skills. Consistent with Foushee's (1984) differentiation of teamwork and task work, McIntyre & Salas (1995) suggested that there are two tracks of skills (task work and teamwork skills) that must be mastered within teams. They argued that task work skills were necessary but insufficient for effective team outcomes and that teamwork skills are required as well. According to Morgan, Glickman,

Woodard, Blaiwes and Salas (1986), task work skills are skills that members must understand and acquire for actual task performance, i.e., the technical expertise and performance strategies. Teamwork skills are the "behavioral" and "attitudinal" responses needed for team members to perform as part of an interdependent team. Others have argued that the relationship between task work skills and team effectiveness is mediated by teamwork skills (Hackman & Morris, 1975; Bass, 1990). These ideas help military team leaders to separate the specific training outcomes when designing team interventions. Moreover, the differentiation of teamwork skills from task work skills gives emphasis to the training of teamwork skills. Teamwork skills can sometimes be confused with the training of technical expertise for team-level operations or given less attention. The differentiation of teamwork and task work was discussed in a research by Mathieu & Rapp (2009) who showed that teams which put in effort to establish the basic foundations of either teamwork or task work emerged more effective over time. However, teams which built the foundation for both teamwork (which they referred to as "team charters") *and* task work (which they referred to as "performance strategies") were more effective than other teams.

Team processes vs. team interventions. Essens, Vogelaar, Mylle, Blendell, Paris, Halpin & Baranski (2005) conducted a comprehensive review of team effectiveness models. They found that the team effectiveness model by Tannenbaum, Beard & Salas (1992) effectively distinguished "team processes" from "team-related interventions" such as individual training, team training and team building which affect team processes. The model also distinguished outcomes relating to "team changes" and "individual changes" from "team performance outcomes". Team changes include new norms, roles, new communications structures and new processes. Team performance is measured in terms of the

Teams need to develop both task work and teamwork skills

Source: Brent Stirton/Reportage/Getty Images

quantity, quality, time, errors, cost and overall productivity. Individual changes include enhanced skills, change in attitudes and motivation. Team and individual changes can influence the team's performance, which in turn serve as feedback on the task and individual characteristics. By differentiating the different types of outcomes, Tannenbaum et al. (1992) provided not only clearer performance variables that may be measured in the military but also highlighted the various team processes that could be changed through team interventions. Today, NATO has its own Command Team Effectiveness Framework (CTEF) based on the team effectiveness model by Tannenbaum et al. (1992).

Command Team Effectiveness Framework (CTEF)

The NATO's Command Team Effectiveness Framework or CTEF (see Figure 7-2 below) is a comprehensive model that emphasizes aspects of teamwork that are directly relevant to most military command teams. The framework uses language that most military leaders would find familiar.

FIGURE 7-2: The CTEF Model

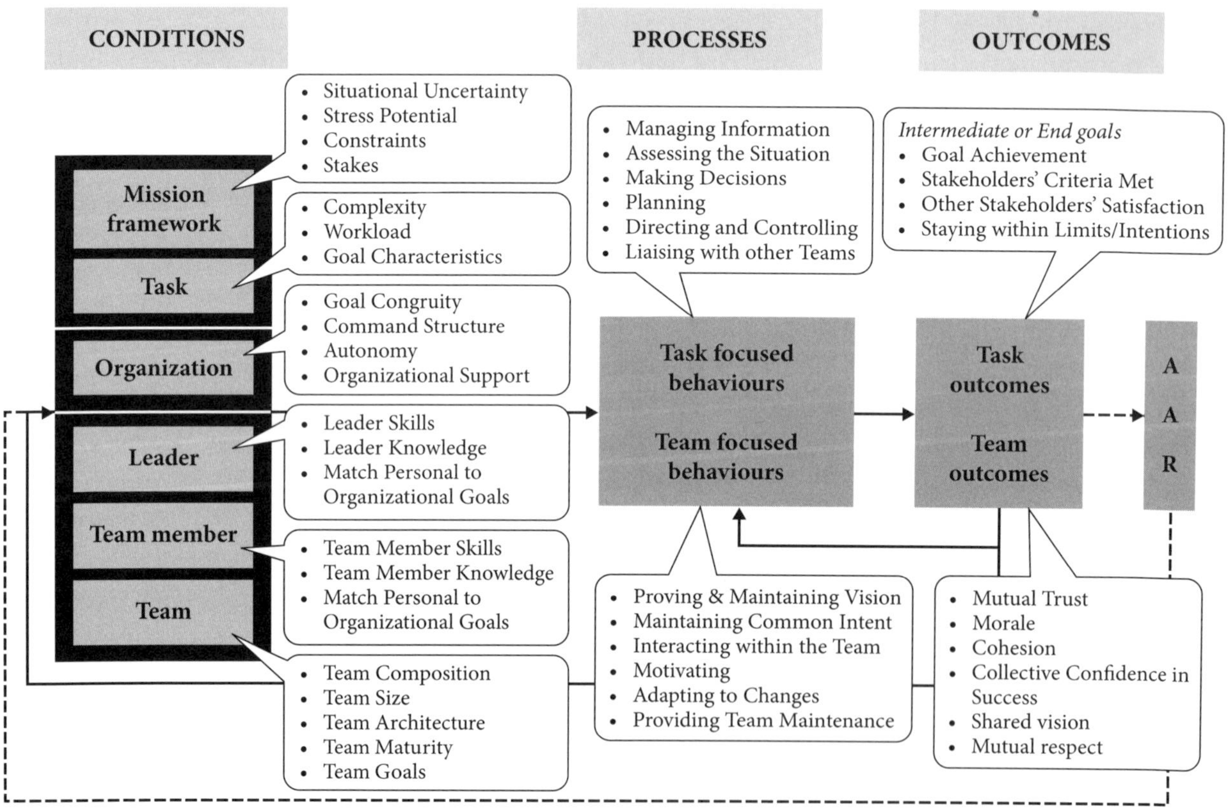

Source: Essens et al., 2005. NATO, RTO

The CTEF instrument, developed and updated by NATO in 2010 after operational applications, serves as a useful diagnostic and feedback tool to improve teamwork in the military. Similar to the feedback processes depicted in Tannanbaum et al.'s (1992) model, the CTEF uses the feedback process to initiate changes that are needed by team leaders at the input stage before the next team performance. The model highlights the leader's skills, knowledge and the importance of aligning one's personal goals to the organization's. What may be unique to the military is the After Action Review (AAR) process that provides additional information to the input stage which military leaders could pay attention to when training and preparing their teams for effective performance in the future.

Leadership Factor in Team Effectiveness Models

As earlier mentioned, Zaccaro et al. (2001) proposed that "team leadership" is one of the three fundamentals of team effectiveness. Team leadership here refers to individuals who are responsible for defining team goals and for developing and structuring the team to accomplish these missions. According to Zaccaro et al. (2001), teamwork is a function of how well the team leader defines "team directions" and organizes the team to "maximize progress along such directions". Zaccaro and Klimoski (2002) argued that "leadership processes serve as one of the most crucial determinants of team effectiveness, particularly in organizations in which leaders are likely to be formally established as the center of key system dynamics". Two other team effectiveness models feature the "leadership" factor in their models and these are the models by Blendell, Henderson, Molloy, & Pascual (2001) and Shanahan (2001). Blendell et al. (2001) included "leadership style" among other variables under input factors and also "leadership" as a process factor in their model. Shanahan (2001) proposed four main elements in team effectiveness: "Process", "Inputs", "Outputs" and "Structure"; "Leadership" together with "task work" and "teamwork" formed a separate cluster of variables under "Process". "Leadership" in these models is treated as a separate function that has to be performed either by an appointed or an emergent team leader and it involves supervision, feedback, goal setting and prioritization and role modeling. While "leadership style" (regarded as a consistent pattern of leadership behaviors) may influence team processes, the more commonly accepted approach in the literature has been "leadership as a process" in team effectiveness.

Zaccaro et al. (2001) described the four processes of how leadership influenced team effectiveness:

- **Through leadership and team cognitive processes.** How leaders make sense of and give sense to the activities may affect the accuracy and degree of sharedness in team mental models. According to Zaccaro et al. (2001), teams will engage in "collective information processing" when leaders "develop members' capabilities for information processing and encourage their participation in team problem solving". In addition, the authors suggested that teams will collectively process information when leaders are able to "develop and motivate member metacognitive capacities and provide appropriate performance feedback". The team leader's role in influencing team cognitive processes is consistent with the idea of shared mental models (SMMs) which will be discussed in Chapter 8.

- **Through leadership and team motivation.** The authors suggested that leaders who set high-performance goals, encourage members to strive for these goals, and show them how to achieve the goals, will have teams with higher team efficacy and cohesion. Leaders must not only set high performance expectations for their teams but also develop their team's capabilities to achieve high performance by supporting them with strategies for effective performance. Teams that learn how to achieve high performance are likely to emerge more confident in their ability to meet performance standards.
- **Through leadership and team affect.** The authors proposed that when leaders provide structure and direction (e.g., clear performance goals, role assignments, and performance strategies), their teams will better manage emotional or stressful circumstances than those whose leaders do not. By helping team members clarify their roles, tasks and what needs to be achieved at the team-level and being clear on the team performance strategy or alternative strategies, leaders can enable their teams to cope effectively when they are faced with performance challenges.
- **Leadership and team coordination processes**. The authors proposed that teams will be better coordinated and more effective when leaders make sure that individual member knowledge, skills and abilities (KSAs) meet their role demands. Leaders must suggest ways to achieve team goals, monitor and provide feedback on performance, and when the team environment changes, take actions to improve team member inputs. Leaders must pay attention to team members' KSAs and how they affect team-level performance so that they can guide team member's training and development to ensure that the team-level requirements are met.

Zaccaro et al. (2001) suggested that in hierarchical organizations like the military, leadership factors account for much of team performance and that a good understanding of the leadership factors will determine how teams and leaders are trained and developed to be effective. In addition, team processes in team-based organizations like the military have a greater and more important influence on leadership and organizational effectiveness.

Team Leadership — the "Functional Approach"

Occasionally in the literature, the term, "team leadership" may be used to refer to the leadership influence of a team in a "shared" or "distributed" sense rather than the influence of an individual appointed to lead a team. The idea of "shared" leadership or "distributed leadership" which will be discussed in Chapter 9 is different from the concept of "team leadership". The more commonly accepted view in the team effectiveness literature is the functional approach to understanding "team leadership" which refers to the leadership provided by an appointed or focal leader in a team. "Team leadership" here differs from the idea of leading individuals through dyadic or interpersonal relationships. The behaviors required of a leader to fulfill the team leadership role are very different from those that "direct" or "passively convey information". Military leaders lead individual soldiers but more often in the military, they lead teams of soldiers. Leading teams is more than about influencing the task outcomes. Team leadership is more than "directing" or "supervising" a team. When leading individual soldiers, the leader needs to establish interpersonal relationships with each of the followers and be concerned with their individual performances. When leading teams, military leaders need to be primarily concerned about the team-level issues while taking into consideration how individual members' development and performances affect team outcomes. What leaders additionally do to add value to a team has been best explained by the functional leadership theory.

Functional Leadership Theory. The functional approach is based on the idea that a leader plays a certain functional role that "adds value" to a team. McGrath's (1962) functional leadership theory asserts that leaders need to "do, or get done, whatever is not being adequately handled for group needs" (cited in Zaccaro et al., 2001). Zaccaro et al. (2001) explained that a leader's role is to "translate the demands and needs of the environment, task, and team into a pattern of leader behaviors that will

Beyond conveying instructions and orders, leaders should also play a functional role that "adds value" to the team.

Source: CyberPioneer, Ministry of Defence, Singapore

enable the team to be successful". Leaders must interpret and define the events in the environment for their team as well as make decisions to solve problems in the team (Zaccaro et al., 2001). Zaccaro, Heinen & Shuffler (2009) reiterated the functional leadership theory, arguing that team leadership must be exercised in order for the team to be effective and team leaders must help members "achieve a synergistic threshold, where collective effort accomplishes more than the sum of individual abilities or efforts". Groups or teams are usually unable to achieve collective effort due to "process loss" (Steiner, 1972; Zaccaro et al., 2009). As discussed earlier, process loss in a team occurs whenever team members are unable to combine their capabilities and pool their efforts together "in a concerted direction" or the "unwillingness or inability of members to exert sufficient levels of individual effort". Zaccaro et al. (2009) suggest that under such circumstances team leaders must step in to "provide the direction for collective action" and help the team to minimize process loss.

The functional leadership theory may explain "what else" leaders need to do for their team. Leaders are often needed to build and develop their teams from their inception. Military leaders have this additional role that they must perform early in the team's formation. For example, platoon or squad leaders who receive new soldiers into their units will need to pay attention to the socialization of individual soldiers into a team context at the earliest stage. A military team leader will need to facilitate the socialization processes during this critical or formative period so that the new soldier is able to build positive relationships with his/her teammates and develop a common identity with the primary group. The U.S. Army Field Manual 22-102 (1987) on forming combat-ready teams, emphasized functions that need to be performed by the leader. At the formative stage, the leader must ensure that a soldier fits well into a team. The doctrine described leader actions such as spending time with new soldiers to ensure the soldier's integration with the primary unit, discussing the soldier's goals and ensuring that the soldier fits into the team. The doctrine also highlighted that the primary role of a leader is to "encourage acceptance, open communication, develop team members' reliance on one another, and promote team acceptance of shared standards and values". It also emphasized that the leader must not only role model how to develop close relationships with the soldiers but also "set and enforce standards . . . in the development of closer relationships".

Leader Behaviors in Teams

Burke, Stagl, Klein, Goodwin, Salas & Halpin (2006) conducted a meta-analysis to understand leader behaviors that were responsible for team outcomes such as "perceived team effectiveness", "team productivity" and "team learning". The Team Leadership Framework by Burke et al. integrated the "taxonomies of leadership functions" proposed by Fleishman, Mumford, Zaccaro, Levin, Korotkin, & Hein (1991) and the "conditions for team effectiveness" proposed by Hackman (2002). (For more on the framework, readers are adviced to refer to Burke et al. (2006).) Fleishman et al.'s (1991) taxonomies include "information search and structure", "information use in problems solving", "managing personnel resource" and "managing material resources" while Hackman's (2002) four conditions for team effectiveness include "having a compelling direction", "an enabling structure", "a supportive organizational context" and "expert coaching" (cited in Burke et al., 2006). The researchers categorized "person-focused" and "task-focused" behaviors under the relevant taxonomies and conditions for team effectiveness and investigated their relationships to the team outcomes.

Person-focused Leadership Behaviors. Burke et al. (2006) found that leadership behaviors that were "person-focused" were strongly related to team learning. These behaviors include the use of (a) "transformational" behaviors where the leader pays attention to both the performance and the developmental aspects of the followers, (b) "consideration" which refers to behaviors that are focused on relationships and cohesion, and (c) "empowerment" behaviors that develop followers' self-management skills. Person-focused behaviors were significantly related to all three team outcomes (perceived team effectiveness, team productivity and team learning). Of these behaviors, empowerment behaviors that were "developmental", and in particular behaviors related to the roles of coaching, monitoring and feedback as well as participative, facilitative, consultative leadership styles accounted for 30% of the variance in team learning. Although the researchers indicate the need for more research on leadership processes related to team learning, the findings from the study suggest that training efforts could focus on developing people-focused behaviors, particularly for "team learning" as an outcome. Team learning is an important concept related to team performance in complex environments (see Chapter 8).

Task-focused Leadership Behaviors. Leadership behaviors related to the task-focused category are "directive" and "autocratic" and include (a) transactional behaviors, for example, offering rewards and praise for good performance, (b) behaviors associated with "initiating structure" which are directly related to the accomplishment of tasks, and (c) "boundary spanning" behaviors, for example, "politically orientated communication" to increase the resources available to the team and "networking communication" to "expand the amount and variety of information" that the team can draw from. The researchers found that task-focused behaviors were moderately related to "perceived team effectiveness" and "team productivity" while boundary-spanning behaviors in particular explained more of the variance in perceived team effectiveness. Based on the findings from the study, team leaders could focus on developing boundary-spanning behaviors to achieve positive team performance outcomes.

Trust in Teams and Trust in Leadership

While trust has always been a central concept in intra-team relations, there has been some increasing interest in how trust is developed in the context of distributed teams in virtual environment as well as in the idea of "swift trust" in emergency teams. Building trust in both teams and the leader is one of the most important things leaders must do. Rousseau, Sitkin, Burt & Camerer (1998) defined trust as a "psychological state comprising the intention to accept vulnerability based upon positive expectations

Leaders play an important role in building trust in teams.
Source: Ojo Images/Workbook Stock/Getty Images

of the intentions or behavior of another" (cited in Burke, Sims, Lazzara & Salas, 2007). According to Adam & Webb (2000) who studied the concept of trust in the military context, most theorists agreed that trust is based on one's expectations about how others like to behave in the future and that trust predicts how one will behave toward others.

Trust in Military Teams. Adams & Webb (2000) argued that there were two types of trust — "person-based trust" and "category-based trust" (see also Adams, Bryant & Webb, 2001). Person-based trust is developed as a result of directly interacting with others. Category-based trust is trust that exists in the absence of direct and personal contact, for example, through shared membership in a group with another person (e.g., we are likely to trust others who have similar cultural understandings as us). Trust in the military team is likely to be influenced by both person and category-based factors, for example, identification within a small team (Adams and Webb, 2000). The authors argued that trust in military teams had several benefits like reducing the need to defensively monitor others who might not behave according to our expectations and the need for formal control measures to guide behavior. The authors also argued that trust facilitated cooperation and open communication that allowed members to share information freely. Trust also decreased conflict in work environments (Simmon & Peterson, 2001; cited in Adams & Webb, 2000). These researchers found that when trust was low, people tended to think that others had ulterior motives. When trust was high, people tended to interpret the ambiguous information as purely task-related rather than believe that others came with hidden agendas (Adams & Webb, 2000). Such attitudes are important in a military team and the findings emphasize the need for military team leaders to consciously build trust and ensure that trust is not easily destroyed.

Trust and Leadership. While it is important that team leaders in the military pay attention to building trust within teams, it is equally important that they build trust in their leadership. In a meta-analysis of trust in leadership, Dirks & Ferrin (2002) found that trust in leadership was significantly related to attitudinal outcomes of job satisfaction, organizational commitment and turnover, behavioral and performance outcomes of altruism, civic virtue, conscientiousness, courtesy and sportsmanship. They developed a framework of trust that included antecedents of trust (e.g. leader actions and practices, follower attributes and relationship attributes) and outcomes of trust (behavioral and performance outcomes, job attitudes and intentions and correlates). In a multi-level review, Burke, Sims, Lazzara, & Salas (2007) developed a model of trust in leadership and approached the concept as: (a) an individual difference, (b) an emergent state or attitude that could develop over time based on the context and which could be both an input as well as a proximal outcome, and, (c) a process through which other behaviors, attitudes and relationships could either be strengthened or weakened. The researchers argued that trust was a key intangible attribute influenced by organizational processes; that trust resulted in important outcomes and within the immediate group, and the qualities of the leader affected trust. In building trust in teams and trust in their leadership, team leaders can approach the subject in the three different ways as suggested in the model by Burke et al. (2007).

Importance of Trust in Military Teams. While it takes a lot of effort on the part of military leaders to build trust in teams and in their leadership, they must also safeguard the trust once it is developed. It is important that military leaders understand that trust is an attribute that can be easily destroyed once it is developed. The survival of a military team in operations is very much dependent on the level of trust that exists amongst the members and in the leader. It is important that military leaders recognise the need to develop trust as early as possible. Trust requires time and is a quality that cannot be achieved overnight.

THE "PRACTICE" OF LEADING TEAMS

Team Building — What Is It?

Team building in the military. The history of team building in the military began with the U.S. Army implementing Organizational Effectiveness (OE) program in 1979 to 1985 after the Vietnam War and when the Army faced complex problems related to the human dimension (Deaner, 1991). Up to 1970s and 1980s, the Organizational Development (OD) efforts such as the OE program tended to assume an external "interventionist" orientation. [OD refers to a set of methodologies and approaches used in a planned effort to systematically improve organizational effectiveness.] The assumption about team building and team development was that one needed "facilitators" unrelated to the team in terms of its membership, to assist the leader to improve the teams and their organization. The U.S. Army trained its line officers holding Captain and Major ranks as internal consultants on a variety of OD techniques including process consultation, team building, survey feedback, conflict resolution and group problem-solving. In 1987, the U.S. Army introduced the Soldier Team Development Manual to formalize or "mainstream" the team development processes systematically within the military (U.S. Army Field Manual 22-102, 1987). The purpose of the field manual was to assist leaders at company level and below in developing soldier teams to meet the challenges of combat. The U.S. Army Field Manual 22-100 (1999) on Army Leadership integrated team development under "improving actions" as part of "direct leadership actions". The stages of team building and leader actions for each stage were also described in the manual.

Technique of Team Building. Team building is the most commonly and widely used Organizational Development (OD) technique in the corporate and business world. In the military, team building programs can easily be confused with cohesion activities which serve a different purpose. In their team effectiveness model, Tannenbaum et al. (1992) distinguished "team building" from "team training" and suggested that team building as an intervention could improve goal setting, team characteristics and interpersonal relationships, to improve team processes. Team building programs typically focus on establishing a shared vision, clarifying functional roles, understanding the interdependence of roles and agreeing to the team rules, goals and action plans. While team leaders in the military are at liberty to adopt any framework proposed by team building or OD experts, it is important is that the team building programs takes into account what has surfaced in team research as being relevant and important to team effectiveness. Figure 7-3 depicts a framework which emphasizes five areas that must ideally be addressed in a team building exercise. The task work dimension involves strategies and plans, and the clarification and understanding of roles and role relationships as well as role interdependence. The teamwork dimension involves the building of trusting and supportive relationships and establishing team rules (which include clarifying acceptable or unacceptable behaviors in a team) and clear communication processes. Shared vision and goals define the team outcomes and these may be described in tangible (performance-related) as well as intangible (team outcomes) forms. Intangible team outcomes include team cohesion, commitment and trust. Team building programs can provide opportunities or the space to address real team issues or problems. When facilitating team building, team leaders must establish a psychologically safe environment for members to discuss difficult and sensitive issues. Activities in a team building program are designed to drive home the importance of team processes such as communication, trust, support and collaboration. There are many team building tools and resources that are easily available to leaders. What team leaders need to bear in mind when facilitating team building programs is the need to have clear objectives (i.e., what needs to be achieved through the program) before deciding on the activities for the team building

FIGURE 7-3. A framework for military team building.

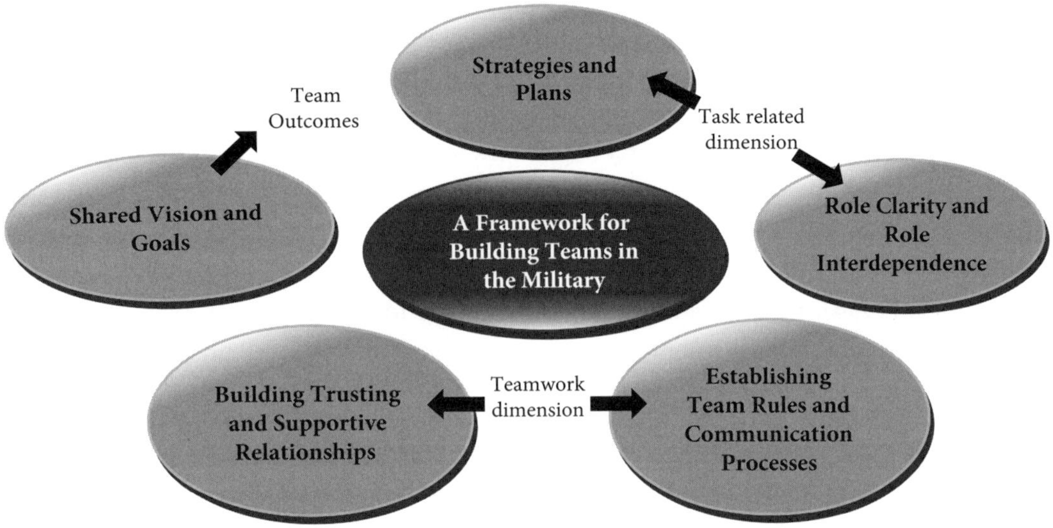

program. Based on what must addressed in a team building exercise (see in Figure 7-3), team leaders will realize that building a team is a continuous process and not an exercise to be accomplished within a day or two. Military team leaders must differentiate team building programs from other military cohesion activities which have a different purpose and which do not share the same processes.

Effectiveness of Team Building. A review by Umstot (1980) found that although team building and other OD efforts were perceived as useful and successful, there was no empirical evidence to suggest that these efforts led to team effectiveness or organizational effectiveness. DeMuese & Liebowitz (1981) in their empirical analysis of team building research found that team building efforts appeared to improve attitudes, perception and behaviors and organizational effectiveness. Due to the lack of rigor in the research methods used to evaluate team building programs, DeMuese & Liebowitz (1981) were unable to conclude what exactly team building programs improved. A review of team effectiveness literature by Kozlowski & Ilgen (2006) also highlighted that the research on team building continued to be weak. Nevertheless, from our understanding about teams and how teams function in the military, leaving teams to chance is a huge risk that most military organizations would prefer not to take. Simply putting together a group of soldiers will not make a team. Directing or supervising a team on what needs to be done is far from what it takes to make teams effective. Knowing what makes a team effective is a good starting point in building effective teams. However, the knowledge about team effectiveness is not the same as knowledge about building and developing effective teams. The key question that military leaders must ask is, "How does one use the technique of team building effectively to achieve positive team outcomes?"

Systematic Team Building. Systematic team building requires team leaders to pay attention to contents, processes and tools or instruments used in a team building intervention. This perspective of team building goes beyond just the technique (as described earlier) to include tools, measurement processes and validation. A model of systematic team building for the military is illustrated in Figure 7-4. In terms of content, team leader and the team must be clear about the objectives to be met. In addition, the team leader needs to select teamwork and task work activities (see Figure 7-3), identify outcomes and decide what to measure in terms of the team outcomes. In terms of the processes, team leaders must plan the event, diagnose problems or measure the team status, design the activities (teamwork and task work related) and facilitate the team building event. According to Woodcock & Francis (1994), team building interventions are ideally preceded by a diagnostic phase to identify specific areas for improvement (p. 1). For the type of tools and instruments to be used, team leaders need to be selective and choose valid instruments rather than what is popularly adopted in the business context. Validated and reliable instruments like the NATO CTEF instrument provide a good guide for what to diagnose (measure and assess) prior to designing team building programs. Assessments using heuristics like team climate surveys can serve as useful pre-post measurement tools. The qualitative tool using the heuristics "Strength-Weaknesses-Opportunities-Threat" or SWOT analysis can be used in addition to a team climate survey. Surveys are particularly useful in team building if they include relevant team effectiveness factors reflected in the team effectiveness models like the CTEF model. A well-designed pre-survey can provide useful information, for example, the

FIGURE 7-4: A model of systematic team building in the military.

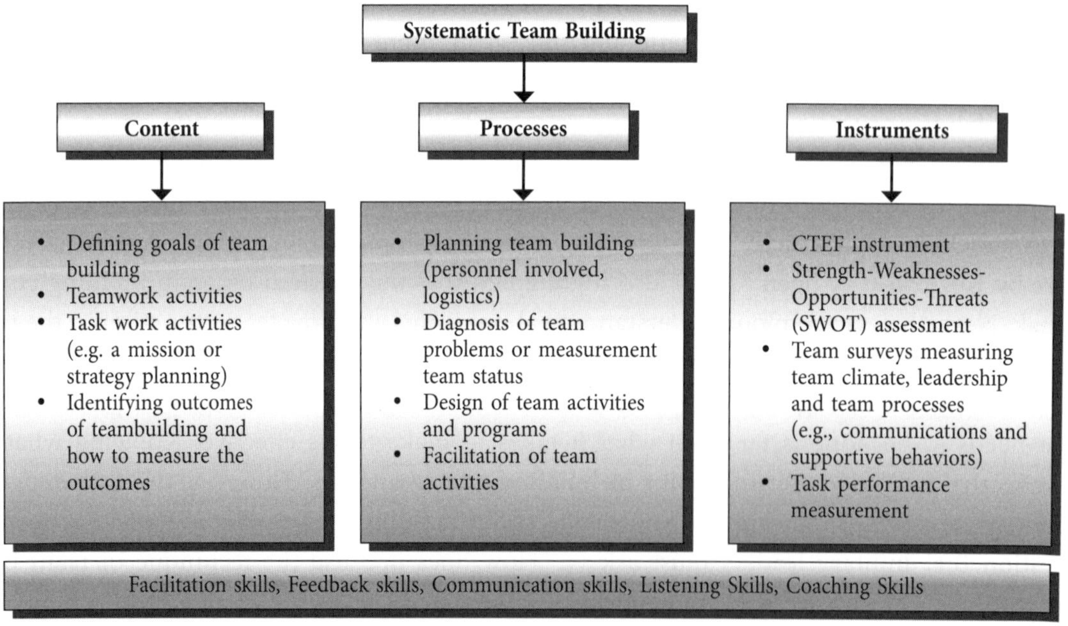

status of the team's climate, and a diagnosis of team problem areas, which can help the team focus on what needs to be achieved in a team building program. A post team building survey could be used to measure the immediate outcomes of the team building program, for example, how clear team members are about the different roles or rules and whether team members identify with the proposed team vision and goals. The post-surveys could also be used to measure the effectiveness of the team building effort and to what extent the objectives of the team building program were met. To support effective team building in the military and especially because team building is a continuous process, team leaders must be equipped with appropriate skills to initiate and implement the team building interventions.

Use of personality tools in team building. Although, personality instruments are popularly used in corporate team building programs, there is currently no empirical research to suggest that personality feedback improves teamwork in the military although some team effectiveness models surface individual differences or personality as one of the input factors. Individual differences and personality factors are usually appropriate if selection methods used in forming a team add value to the team effectiveness. In the military, teams are usually not formed based on individual differences but through other selection methods that include ability or aptitude tests and programs designed for the purpose of selection. Some authors such as Wolf (1988) caution against team building that deals with interpersonal differences and propose that the team intervention focus on defining the mission, developing goals, setting objectives and charting responsibilities. There may be an over-emphasis and over-use of personality instruments in corporate team building programs which if adopted by the military may distract military teams from the real team issues. Personality feedback and other human resource development programs have their limitations in how much change can be expected to have any significant impact on teams and teamwork. There is a need for more research on how personality feedback or development can directly improve team effectiveness. Popular methods of team building programs which last for a day or two may leave teams feeling good immediately after a program but resolving real team issues requires the science, which means conducting proper diagnosis of team issues and developing customized interventions which must be followed through and evaluated over the lifespan of a team. Military forces that take teams and teamwork seriously would need to rely on the science rather than depend on commercialized tools for business use that may not serve military needs. Effective team building programs also require the team leader's involvement and commitment which means that military forces must also pay attention to training leaders in this area.

Team Development — What Must a Leader Do?

As earlier discussed, team building as a team intervention involves the active involvement and facilitation by the team leader. "Team development" is another form of intervention that requires the team leader to play an active role. Team development is a process that team members can initiate to develop "effective social structures and work processes" to improve team processes when the team matures and becomes more competent (Kozlowski & Ilgen, 2006). However, leaving such processes to occur naturally in a team would mean leaving them to chance. Team leaders may need to step in to initiate such processes if team members do not. What models or theories can team leaders use for team development?

Tuckman's Stage Theory. A team development model that is popularly used in the practice of team building and team development is the Stage Theory of Group Development by Bruce Tuckman (1965). It is interesting to note this model has its origins in the military. Tuckman served as a research psychologist with the Naval Medical Research Institute and studied small groups and organizational behavior (Bonebright, 2010). His study was driven by the U.S. Navy's interest in how small groups functioned, in preparation for the Navy's deployment of teams in small crew vessels and stations. Bonebright (2010) reviewed Tuckman's work and highlighted the limitations that Tuckman himself raised in his 1965 article. One of the limitations in the proposed model was the lack of representativeness of the small groups that he studied. His sample of groups was taken from the psychotherapy and mental health domain and was not representative of all groups. The other problem that he raised was the qualitative rather than quantitative analysis which he employed in this study. One study conducted in a classroom setting supported the stage theory of team development and that there was a sequence to team development described by Tuckman. Another study found that it was difficult to identify the stages of development but observed that groups generally tended to develop in a predictable sequence like what Tuckman described and subsequent group development models also presented a similar pattern for group development (Bonebright, 2010). Despite the limitations cited by Tuckman himself and other researchers, the stage model of team development continues to provide a useful framework which team leaders could use to discuss team development. Based on the stage theory, authors like Chapman have prescribed different roles that team leaders could play to facilitate the team processes in the various stages of their team's development. The stage theory has since been modified, integrated and applied in the practice of team leadership. Chapman (1995–2010), for example, integrated the Situational Leadership model with Tuckman's theory and suggested how leaders could play different roles at the different stages of team development (see below).

- **Forming stage**. The forming stage is a stage where the team relies on the leader to guide and direct the team. Members are usually readily agreeable and seek to establish relationships. However, this is a stage where the roles of members are unclear and where the members may have many questions about the team's purpose and objectives. Members may be more focused on finding answers and in trying to warm up to one another in the team. The leader is expected to direct the team, and clarify the purpose and objectives with members. How well a leader plays the role at this stage will determine how successful the team will be in avoiding the problems in the next stage of development.

- **Storming stage.** The storming stage is characterized by disagreements and conflicts. The stage becomes worse when the leader does not do enough to clarify the purpose and objectives for the team. Members may attempt to understand each other better or may form negative perceptions of one another that may hinder teamwork and collaboration. Although the team may become clearer about its role, the team may also face uncertainties. If disagreements and conflicts are not carefully managed, the team may risk forming cliques with team members competing against each other. The leader needs to coach the team to focus on its goals and to self-manage interpersonal issues. If the leader can help the team to overcome the problems typically faced in this stage, the team can then begin to normalise its culture and strengthen its relationships to be better prepared for the challenges associated with the performance stage.
- **Norming stage.** The norming stage is characterized by agreement and consensus and the team responds positively to leader actions. Team members will have clearer roles and responsibilities and become more committed. The team may engage in activities that build team cohesion. The team may also review its processes and team culture. The leader is respected and members may volunteer themselves and share more responsibilities for the team. The leader does well if he is facilitative and enables the team to initiate changes where necessary.
- **Performing stage.** The performing stage is a stage where the leader can afford to intervene less and take on the role of delegating and overseeing the team. The team at this stage is mature, and is more "strategically aware" of its role in the organization. The team has a clear purpose and a shared vision and there is little need for the leader to propose any changes in these aspects. Conflicts are resolved positively and the team may initiate changes in terms of processes and structure. The team is able to focus on both task and relationship issues and to sustain itself. Successful team performances and positive team outcomes will serve to further strengthen team identity and enhance team cohesion. The leader, in addition, could help the team take stock and review performance problems and use the performance opportunities to focus on team learning.
- **Adjourning stage.** The adjourning stage is where a team is dissolved after the mission or task is completed and the purpose fulfilled. Team members look forward to new or different things and if the team is successful, members would move on with a sense of pride and achievement. They can carry forward their team experiences and learning to other team environments. Leaders can help by providing recognition of individual efforts and by being sensitive to how the change affects members in the organization. This stage is typically experienced by most military teams where changes to the team composition and even team structure are expected following mission accomplishments or when there are leadership transitions. In the case of leadership transitions, the outgoing team leader has the responsibility to ensure a smooth transition and handover of duties to the next leader.

Applying Tuckman's Theory in the Military. Tuckman's theory provides a basis for team interventions at various stages in the life of a military team. Team building in the military is to be ideally conducted before the team transits into the storming stage. During the team building, team leaders can focus their team members on a common vision as well as the mission, help them clarify their tasks or functional roles, help members become aware of task interdependencies and encourage members to articulate the desired behaviors in the team. These processes will help team members communicate better, prevent misunderstandings and problems that characterise a storming stage. Once the team building efforts have served to build the foundation of effective military teams, team leaders can turn to team development or team training approaches to train team members and their team in competencies important for performance. While team building efforts are important in the formative stages of teams, the current team effectiveness research suggests that military teams need to go beyond the team building to prepare themselves adequately for challenges in dynamic and stressful environments. The stage theory and the roles team leaders can play in each stage is a relevant and valid approach to team development in the military. Team training approaches will be discussed in the next chapter.

Leader Skills for Team Building and Team Development. What skills or competencies will a leader need in order to build teams effectively? It is always useful to have a trained coach or facilitator to assist leaders in team building but not all military leaders, especially those at the squad, platoon and company levels, have access to such assistance. A closer look at what skills or competencies are required for leaders to build teams suggests that the skills may not be very different from the skills that are required of leaders to conduct many other activities in the military. To support effective team building in the military and especially because team building is a continuous process, team leaders must be equipped with facilitation, feedback, communication, listening and coaching skills. These same skills are required of leaders to carry out other leadership functions. For example, leaders need to have effective feedback skills to conduct performance feedback, coaching skills for guiding individual soldiers' task performances and facilitation skills for conducting regular meetings or other activities with different military teams or personnel. These topics are beyond the scope of this chapter and interested readers may refer to other sources in the OD or training development literature for more information on these skills. For example, Jones and Bearley (2001) documented useful and practical ideas on how to facilitate team development. Military team leaders could stand to benefit a great deal from these generic or enabling skills training if such skills training could be introduced systematically like other independent basic military courses or if these skills training could be integrated into existing military courses or developmental programs that allow military leaders to practice the skills consistently and regularly.

SUMMARY

The key ideas to take away from this chapter are that teams are more than just mere social groupings; that teams cannot be left to chance and leaders play an important role in making teams effective; that it takes systematic effort to build and maintain effective military teams. Team leaders in the military will need a good understanding of team effectiveness, the factors that affect it and how they, as leaders, can build and develop effective military teams.

In the first section of the chapter, the basic concepts about teams and team effectiveness were explained. The chapter also introduced the I-P-O approach and the NATO-CTEF team effectiveness model that provide leaders with a comprehensive framework about team effectiveness in the military context. The differences in the terms "task work" and "teamwork" were highlighted. Teamwork is more complex and involves more than just interpersonal relationships in a team. Leading individual soldiers is very different from leading teams which requires a more involved approach that can be best understood from a functional leadership perspective. The idea of "team leadership" was differentiated from other forms of leadership (i.e., team leadership is not "shared" or "distributed" leadership). "Team leadership" based on the functional leadership theory requires leaders to play a role or function that the team itself (or members) cannot perform. Leadership functions or roles in a team do not inform one how well to perform them. There are effective leader behaviors that must be attended to when leading a team. Person-focused and task-focused behaviors both make a difference to team outcomes but person-focused behaviors, in particular empowerment, make a significant difference to team learning. The chapter also discussed the topic of trust and highlighted the important role leaders have in building trust in teams and trust in their leadership. The ideas and theories from the first section of the chapter have provided the scientific basis for the practice of leading teams in the military.

The second section of the chapter described a widely used OD technique referred to as "team building", a team intervention to improve team effectiveness. Team building involves establishing common or shared goals and plans, improving interpersonal relationships, clarifying roles and agreeing to the team rules, goals and action plans. The section also introduced a team building framework based on the key ideas in the I-P-O approach to team effectiveness. The framework suggests a comprehensive approach to building effective foundations for team performance in the military. The section also presented a model of systematic team building in the military highlighting the contents, processes and instruments for effective team building. The model emphasizes the "science" in the practice by drawing attention to the diagnosis and measurement processes that are needed in team building. Team building which is ideally conducted for teams in their early stages is quite different from "team development". Team development is largely initiated by a more competent and mature team. Team leaders in the military can be actively involved in team development by paying attention to the different roles that are required of the leader at the various stages of team development. Team leaders need skills like effective feedback, facilitation and coaching skills to build teams. Teams in the military, however, need more than just team building or team development to prepare for 21st century challenges as will be discussed in the next chapter on preparing and training teams for complex environments. The skills and behaviors required of leaders for preparing and training teams for 21st century operations are also different as will be discussed in the next chapter.

Key Concepts

- Command Team Effectiveness Framework (CTEF)
- Functional Leadership Theory
- Group vs Team
- Input-Process-Output (I-P-O) model
- Person-focused leadership behaviors
- Process Loss
- Process Gain
- Task-focused leadership behaviors
- Task outcomes
- Team outcomes
- Task work
- Team development
- Team effectiveness
- Team leadership
- Team training
- Teamwork
- Tuckman's theory
- Systematic team building

Questions

- What are your experiences leading teams? How would you explain your experiences based on what you have learnt about "team effectiveness"?
- Reflect on leader behaviors (positive and negative) you have experienced in a team. The experiences can be your own or someone else you had a chance to observe. How would you describe these behaviors from the functional perspective of "team leadership"?
- What have you learnt about "systematic team building" that you find particularly useful for your application? What factors would you consider when building a team in your unit?

References

Adams, B.D., & Webb, R.D.G. (2000). Trust in Small Military Teams. Retrieved from: http://www.dodccrp.org/events/7th_ICCRTS/Tracks/pdf/006.PDF.

Adams, B.D., Bryant, D.J., & Webb, R.D.G (2001). *Trust in Teams: Literature Review* (Technical Report DCIEM No. CR-2001-042). Canada: Department of National Defense.

Bass, B.M. (1990). *Bass & Stogdill's handbook of leadership* (3rd ed.) New York: The Free Press.

Blendell, C., Henderson, S.M., Molloy, J.J., & Pascual, R.G. (2001). *Team performance shaping factors in Integrated Performance Modeling Environment* (Unpublished DERA report). Fort Halstead, UK: DERA.

Bonebright, D.A. (2010). Forty years of storming: a historical review of Tuckman's model of small group development. *Human Resource Development International, 13(1),* 111–120.

Burke, C.S., Stagel, K.C., Klein, C., Goodwin, G.F., Salas, E., & Halpin, S.M. (2006). What type of leadership behaviors are functional in teams? A meta-analysis. *The Leadership Quarterly, 17,* 288-307.

Burke, S., Sims, D., Lazzara, E., & Salas, E. (2007). Trust in leadership: A multi-level review and integration. *The Leadership Quarterly, 2007, 18,* 606–632.

Cannon-Bowers, J.A., & Salas, E. (1998). *Making Decisions Under Stress: Implications for individual and team training.* Washington, DC: American Psychological Association.

Chapman, A. (1995–2010). Tuckman's Forming Storming Norming Performing Model. Businessballs.com. Retrieved from: www.businessballs.com/tuckmanformingstormingnormingperforming.htm.

Deaner, C.M.D. (1991). The U.S. Army Organizational Effectiveness Program: A Eulogy and Critique. *Public Administration Quarterly, 15(1),* 12–31.

DeChurch, L.A., & Marks, M.A. (2006). Leadership in Multiteam Systems. *Journal of Applied Psychology, 9(2),* 311–329.

DeMuese, K.P., & Liebowitz, S.J. (1981). An empirical analysis of team building research. *Group & Organisation Studies, 6(3),* 357–378.

Dirks, K.T., & Ferrin, D.L. (2002). Trust in Leadership: Meta-analytic findings and implications for research and practice. *Journal of Applied Psychology, 87(4),* 611–628.

Driskell, J.E., Salas, E., & Hogan, R. (1987). *A taxonomy for composing effective naval teams* (Technical Report Code 712). Naval Training Systems Center: Orlando, FL.

Essens, P., Vogelaar, A., Mylle, J., Blendell, C., Paris, C., Halpin, S., & Baranski, J. (2005). *Military Command Team Effectiveness: Model and Instrument for Assessment and Improvement* (RTO Technical Report, TR-HFM-087). NATO, RTO.

Fleishman, E.A., Mumford, M.D., Zaccaro, S.J., Levin, K.Y., Korotkin, A.L., & Hein, M.B. (1991). Taxonomic efforts in the description of leader behavior: a synthesis and functional interpretation. *Leadership Quarterly, 2(4),* 245–287.

Foushee, H.C. (1984). Dyads and triads at 35,000 feet: Factors affecting group processes and aircrew performance. *American Psychologist, 39,* 885–893.

Hackman, J.R. (2002). *Leading Teams: Setting the stage for great performances.* Boston: HBS Press.

Hackman, J.R., & Morris, C.G. (1975). Group tasks, group interaction process, and group performance effectiveness: A review and proposed integration. *Advances in Experimental Social Psychology, 8,* 45–99.

Hughs, R.L., Ginnett, R.C., & Curphy, G.J. (2002). *Leadership: Enhancing The Lessons of Experience.* New York: McGraw-Hill.

Jones, J.E., & Bearley, W.L. (2001). Facilitating team development: A view from the field. *Group Facilitation, 3,* 56–64.

Katz, D., & Kahn, R.L. (1978). *The social psychology of organizations.* New York, NY: Wiley.

Kozlowski, S.W.J., & Ilgen, D.R. (2006). Enhancing the Effectiveness of Work Groups and Teams. *Psychological Science in the Public Interest, 7,* 77–124.

Mathieu, J.E., & Rapp, T.L. (2009). Laying the foundation for successful team performance trajectories: The roles of team charter and performance strategies. *Journal of Applied Psychology, 94(1),* 90–103.

McGrath, J.E. (1962). Leadership behavior: Some requirements for leadership training. Washington, DC: U.S. Civil Service Commission.

McGrath, I.E. (1964). *Social Psychology: A Brief Introduction*. New York: Holt.

McIntyre, M.G., & Salas, E. (1995). Measuring and managing for team performance: Emerging principles from complex environments. In R.A. Guzzo, E. Salas & Associates (Eds.), *Team effectiveness and decision making in organizations*. South Florida: Jossey-Bass Publishers.

Morgan, B.B. Jr., Glickman, A.S., Woodard, E.A., Blaiwes, A.S., & Salas, E. (1986). *Measurement of team behaviors in a Navy environment.* (Technical report NTSC-TR-86-014). Orlando, FL: Naval Training Systems Center.

Prince, C., & Salas, E. (1993). Training and research for teamwork in the military aircrew. In E.L. Wiener, B.G. Kanki, & R.L. Helmreich (Eds.), *Cockpit Resource Management* (pp. 337–366). Orlando, FL: Academic Press.

Rousseau, D.M., Sitkin, S.B., Burt, R.S., & Camerer, C. (1998). Not so different after all: a cross-discipline view of trust. *Academy of Management Review, 23*, 393–404.

Salas, E., Bowers, C.A., & Cannon-Bowers, J.A. (1995). Military Team Research: 10 years of progress. *Military Psychology, 7(2)*, 55–75.

Salas, E., Cannon-Bowers, J.A., Church-Payne, S., & Smith-Jentsch, K.A. (1998). Teams and Teamwork in the military. In C. Cronin (Ed.), *Military Psychology: An Introduction* (pp. 71–87). Upper Saddle River, NJ: Pearson.

Salas, E., Dickinson, T.L., Converse, S., & Tannenbaum, S.I. (1992). Toward an understanding of team performance and training. In R. W. Swezey, & E. Salas (Eds.), *Teams: their training and performance*, (pp. 3–29). Norwood, NJ: Ablex Publishing.

Salas, E., Rosen, M.A., Burke, C.S., & Goodwin, G.F. (2009). "The wisdom of collectives in organizations: An update of the teamwork competencies. In E. Salas, G. F. Goodwin & C. S. Burke (Eds.), *Team Effectiveness in Complex Organisations: Cross-Disciplinary Perspectives and Approaches*. New York: Taylor & Francis Group.

Salas, E., Wilson, K.A, Burke, C.S., Wightman, D.C., & Howse, W.R. (2006). Crew Resource Management Training: Research, Practice, and Lessons Learned. In Robert C. Williges (Ed.), *Reviews of Human Factors and Ergonomics, 2*. Santa Monica, CA: Human Factors and Ergonomics Society.

Shanahan, P. (2001). *Mapping team performance shaping factors*. Quineti: Fort Halstead.

Simons, T., & Peterson, R. (2000). Task conflict and relationship conflict in top management teams: The pivotal role of intragroup trust. *Journal of Applied Psychology, 85(1)*, 102–111.

Steiner, I. (1972). *Group process and productivity.* New York: Academic Press.

Stouffer, S.A., Suchman, E.A., DeVinney, L.C., Star, S.A., & Williams, R.M. (1949). *Studies in Social Psychology in World War II: The American Soldier Vols. 1–4*. Princeton: Princeton University Press.

Tannenbaum, S.I., Beard, R.L., & Salas, E. (1992). Team building and its influence on team effectiveness: An examination of conceptual and empirical developments. In K. Kelley (Ed.), *Issues, theory, and research in industrial/organizational psychology* (pp. 117–153). New York: Elsevier Science.

Tuckman, B.W. (1965). Developmental sequence in small groups. *Psychological Bulletin, 63(6)*, 384–399.

Umstot, D. (1980). Organization Development Technology And the Military: A Surprising Merger? *Academy of Management Review 1980, 5(2)*, 189–201.

U.S. Army. (1987). *Soldier Team Development* (FM 22-102). Washington, DC: Headquarters, Department of the Army.

U.S. Army. (1999). *Army Leadership* (FM 22-100). Washington, DC: Headquarters, Department of the Army.

Wolf, M.F. (1988). Before you try team building. *Research Technology Management, 31(1)*, 6–8.

Woodcock, M., & Francis, D. (1994). *Teambuilding Strategy*. Hamshire, England: Gower Publishing.

Zaccaro, S.J., Hienen, B., & Shuffler, M. (2009). Team Leadership and Team Effectiveness. In E. Salas, G.F. Goodwin & C.S. Burke (Eds.), *Team Effectiveness in Complex Organisations: Cross Disciplinary Perspectives and Approaches*. New York: Taylor & Francis Group.

Zaccaro, S.J., & Klimoski, R. (2002). The Interface of Leadership and Team Processes. *Group & Organization Management, 27(1)*, 4–13.

Zaccaro, S.J., Rittman, A.L., & Marks, M.A. (2001). Team Leadership. *The Leadership Quarterly, 12*, 451–483.

8

Leading Military Teams for Adaptive Performance in Complex Environments

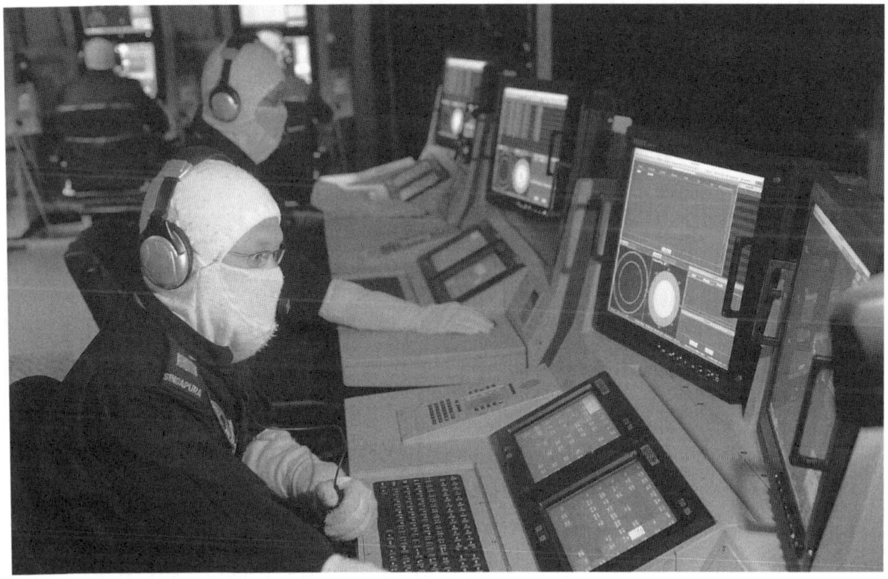

Source: CyberPioneer, Ministry of Defence, Singapore

OVERVIEW

Building on the foundational concepts covered in Chapter 7 on the science and practice of leading military teams, this chapter introduces a new of body knowledge that focuses on (a) team performance and team competencies required in complex environments typical of 21st century military operations and (b) the role of military leaders in training and preparing teams for such environments. The topic of leading military teams in complex environments is in itself highly advanced because of the many concepts and ideas that have been integrated from different fields in industrial-organizational psychology (teams, teamwork, leadership), organizational development and even cognitive psychology. This chapter features team research conducted as part of the U.S. Navy-sponsored team research program known as "Tactical Decision Making Under Stress" (TADMUS) which was initiated following an incident in 1988 where an AEGIS cruiser, the USS Vincennes, accidentally shot down a civilian airliner. The TADMUS program was a major attempt to understand how to train military teams and prepare them for complex environments. Various theories, conceptual models and team training strategies developed by behavioral scientists and tested for application in the U.S. Navy are introduced in this chapter. This chapter will also provide military leaders with a good understanding of what it takes to train and prepare teams for 21st century operations, highlighting specific processes and leadership behaviors critical for the development of adaptive, learning military teams.

TEAM PERFORMANCE CHALLENGES IN COMPLEX SETTINGS

Cognitive, Behavioral and Affective Demands

Chapter 2 provided a description of the key challenges facing individual soldiers or leaders in 21st century military operations. The conditions in the new environment have also changed the way military teams function. At the heart of the challenge is how teams communicate and coordinate tasks under dynamic, high-stress, high-demand environments. Besides distracting team members from focusing on tasks and team processes, stress from the operating environment also increases the chances of errors that could lead to disastrous consequences like the USS Vincennes incident (see Box 8-1 for a short discussion of the incident). Preventing errors and avoiding consequences like that of the USS Vincennes incident is, therefore, one major source of stress for military teams in 21st century operations.

It is important to first understand the kinds of demands placed on teams in such complex environments. As illustrated in Figure 8-1, the demands can be associated with the cognitive, behavioral or affective dimensions. Cognitive complexities arise from tasks that require accurate processing of information under time pressure. For example, computer-guided weapon systems generating huge amounts of data in a command information center (CIC) is a typical scenario in complex military operations where information needs to be processed and interpreted for decision making under compressed time. The team in this instance has to rely on its members to provide accurate inputs to make critical decisions and team members processing the data can experience high levels of cognitive overload and stress. The data could appear confusing, contradictory and highly complex especially under dynamic conditions. Even though there are advanced technologies such as the Group Decision Support Systems (GDSS) to help military personnel solve complex problems and minimize errors,

BOX 8-1

USS Vincennes Incident and the TADMUS program

On July 3, 1988 an AEGIS cruiser, the USS Vincennes, accidentally shot down a commercial aircraft, Flight 655, over the Arabian Gulf and killed all 290 people on board. The crew of the USS Vincennes misidentified the Iranian Airbus A300 as an Iranian Air Force F-14. How did the crew make such a mistake? The answer was not so simple. Investigations revealed that there may be several factors that caused the crew to believe that the Airbus was a hostile aircraft. The tendency to focus on the crew error is normal but most often in incidents like the USS Vincennes, the tendency to ignore the "situation" or the "context" in which mistakes are made are real. Some argued that the mistake was a result of the political climate surrounding the incident which influenced the crew's orientation and decision making process. The quality of the crew and how prepared they were for their mission are some other factors that might explain the crew's reaction to the demands of the situation.

As the US Navy also experienced other Navy incidents around the same time, it had questions about its own training and decision support systems. Following the USS Vincennes incident, the US Office of Naval Research commissioned the "Tactical Decision Making Under Stress" or TADMUS research program which was conducted throughout the 1990s. The program generated over 250 research papers and the edited book entitled, *Making Decisions Under Stress* by Canon-Bowers & Salas (1998). The researchers in TADMUS used relevant academic theories and models to understand how teams made decisions under stress. They carried out field studies and research that were conducted using realistic combat scenarios like the Navy ship's Combat Information Center (CIC). The research from TADMUS studies has since created unique knowledge about teams in the military which military forces could consider for implementation as part of their systematic training and development of military teams and team leaders. For detailed studies in the TADMUS programs, readers are adviced to refer to the book, *Making Decisions under Stress: Implications for Individual and Team Training.*

The USS Vincennes

Relatives of victims of Iranian airliners at a ceremony marking the 15th year of the shooting in 2003.

Source: Time & Life Pictures/Getty Images (left) and AFP/Getty Images (right)

FIGURE 8-1: Team Demands in the 21st century

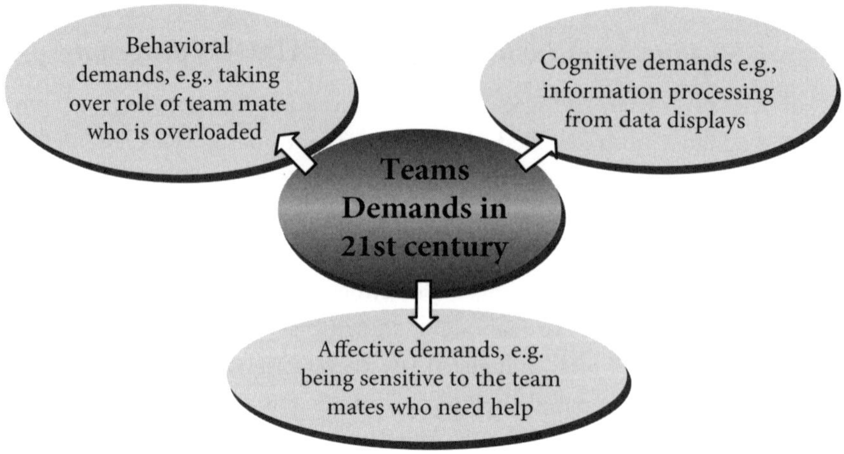

military organizations still need to rely on well-trained individuals in teams to interpret information accurately and quickly under stress and feed the information for critical team decisions. Cognitive demands could also come from the need to manage or minimise potential errors in addition to making critical decisions under stress. Building cognitive capabilities such as critical thinking skills to enhance decision making under stress are essential for teams in complex environments.

Highly interdependent military teams will need to deal with and manage the dysfunctions of individual team members and poor team interactions. Under stress, individuals in teams can become too focused on their individual tasks, become less sensitive to the needs of other team members and lose the team perspective (Driskell & Johnston, 1998). Team members will need to share information and know what to do in order to support other members. When dealing with new task situations, teams need to learn skills and behaviors that enable them to be proactive, flexible and adaptive. Teams, therefore, need specialized team training to recognize, prepare and manage team challenges and to ensure that team processes support rather than undermine the team's mission. Team training should focus on the behavioral and affective dimensions over and above the cognitive dimension. Such training will be necessary in order for military organizations to function like "high-reliability organizations" — a term used in the Canadian Forces' Leadership Manual (2005, p. 28) to refer to organizations where the consequence of errors that are potentially considered "catastrophic" — much like what happened in the USS Vincennes incident.

Implications for Team Leaders

What are the implications, therefore, for team leaders in complex military operations? Figure 8-2 illustrates the four knowledge or skill components that team leaders will need in order to prepare for complex operations. Firstly, leaders need to know what is demanded of teams in complex settings. They need to interpret the environment for their teams, analyze situations and their implications, and communicate effectively under stress. They also need to facilitate and pull individual efforts together and enable critical team processes that can be affected by stress. It is, therefore, imperative that military leaders possess a good understanding and awareness of the unique demands placed on teams in complex military operations. Secondly, leaders must have a good understanding of both individual-level competencies as well as team-level competencies that are critical for adaptive team performance. Team leaders are instrumental in developing these competencies in their teams. Thirdly, team leaders need to do more than team building and cohesion activities. They need to be equipped with training strategies to prepare teams for complex environments. Lastly, team leaders also need to learn specific behaviors and skills to support team learning. As it is impossible for team leaders to physically monitor and rectify teamwork processes, team leaders must build team learning capabilities that enable their teams to self-correct their own team processes.

FIGURE 8-2. Knowledge and skills for leading military teams in complex environments.

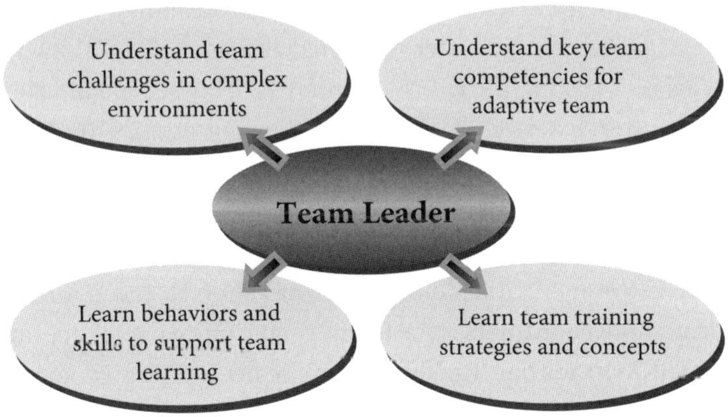

THE TEAM TRAINING APPROACH

It is important that leaders know the different types of team interventions (see Figure 8-3) and what approach to take in order to prepare their teams for complex, 21st century military operations. In a literature review on effective work teams, Kozlowski & Ilgen (2006) differentiated the term, "team training" from the popular OD techniques of "team building" and "team development" which were discussed in Chapter 7. While the objectives of team building and team development programs are to build strong foundations for team effectiveness, team training refers to training that is aimed at developing or training specific competencies such as critical thinking skills in teams through individual-level training or team-level training strategies. The "team training" intervention is a formal

FIGURE 8-3. Types of Team Interventions.

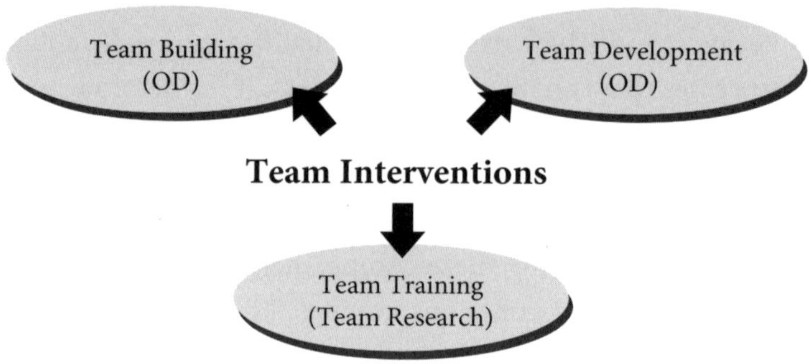

training approach that is systematically conducted and supported with measurements of task work, team processes and outcomes. The training of team competencies at the individual-level is ultimately to help individuals become effective and competent in a team.

The concept of "team training" originated from Cockpit Resource Management training in the 1970s. Cockpit Resource Management was later redefined and labeled as "Crew Resource Management" (CRM) in the 1980s (Salas, Wilson, Burke, Wightman & Howse, et al., 2006). Cockpit Resource Management was created after a series of airline disasters in the 1970s badly affected the airline industry. Lauber (1984) invented the term, Cockpit Resource Management, to refer to the "use of all available resources in terms of information, equipment and people to achieve safe and efficient flight operations" (Lauber, 1984; Salas, et al., 2006). In 1979, the commercial aviation industry was formally introduced to the idea of CRM at a NASA-sponsored workshop to teach aircrews to use all available resources to communicate and coordinate as a team. Some navies refer to such training as "Bridge Resource Management" (BRM) training. In general, any team training program that focuses on team performance and which uses training methods, techniques and measurement instruments tested and validated for effective team performance can be regarded as "team training".

Team training stretches individuals in teams beyond their technical or task competencies to focus on improving team processes such as coordination, communication and support behaviors necessary for effective task performance. Since these team behaviors are identified as critical for team performance under extreme conditions, team leaders in the military need to consciously build these team processes in a structured and systematic way. To do this, military team leaders need to learn the training principles and approaches that are known to be effective for team training. One key principle being proposed for team training is that training must eventually be conducted under simulated task or actual performance conditions.

Results from team research studies thus far support the implementation of team training interventions to enhance team performance under complex environments. Team training programs in the military, however, need to be rationalized together with team building efforts (that are necessary at the formative stages of a team) and the ongoing team development programs, both of which are largely left to team leaders to initiate. The implementation of different team interventions (team training, team building and team development) requires systematic approach which should also include the formal training of leaders to implement the different team interventions. In particular, team leaders will need specialized training to implement team training strategies to prepare and develop their teams in complex environments.

KEY COMPETENCIES FOR TEAMS IN COMPLEX ENVIRONMENTS

The team research literature surfaces several competencies that are important for effective team performance in complex environments. Since the USS Vincennes incident, team researchers have been studying how the competencies work in teams and how to develop training strategies to equip teams with these competencies. Some of the key competencies that team leaders must know are "adaptability" in teams, "critical thinking skills" and "shared mental models" (SMMs). Figure 8-4 shows key team competencies for adaptive team performance in complex environments. Critical thinking skills will also be discussed in the context of team decision making which need to be discussed together with the type of decision approach adopted by the military.

FIGURE 8-4. Critical competencies for adaptive team performance in complex environments.

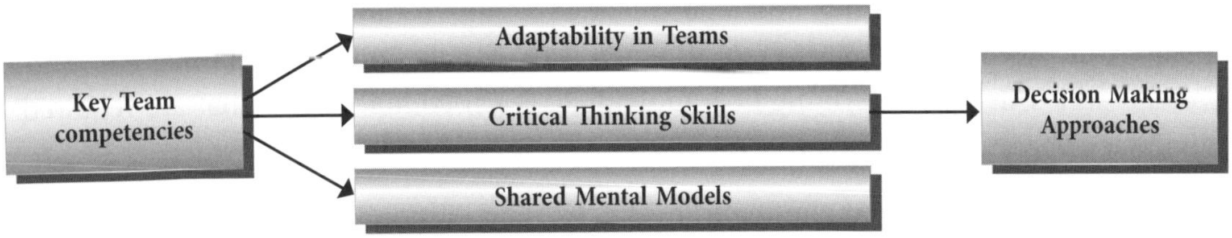

Adaptability in Teams

Developing adaptive expertise in individuals. It is critically important to be able to respond adaptively to team challenges in complex and volatile environments. Two key questions that will be discussed are, "what makes a team adaptive?" and "how to develop adaptability in teams?" One theory about how teams can be developed to adapt and cope in high-stress environments is Kozlowski's (1998) theory of developing "adaptive expertise" which involves the training of individuals to acquire

Military teams need adaptive expertise to meet the demands of 21st century operations.
Source: Scott Nelson/Reportage/Getty Images

two different sets of skills. Kozlowski labeled these as "meta-recognition" and "self-regulatory" skills. He described "meta-recognition skills" as higher level processes that involved "knowledge, awareness, and control of cognitive activity involved in goal attainment". "Self-regulation processes" are processes that involved "planning, monitoring, and adjustment of cognitive and behavioral strategies" needed to achieve sub-goals. According to Kozlowski, the two sets of skills influence "motivation" and "affect" in individuals which in turn determine how individuals learn complex tasks that are required in 21st century military operations.

Developing adaptive expertise at the team level. Kozlowski argued that when individuals applied the skills of meta-recognition and self–regulation beyond the "self", there would be greater awareness of the team situation and what was expected of the individual in a team situation. The awareness would then enable important team processes like team coordination. To achieve adaptive expertise at the team level, team members will need to attend to how they perform their "interdependent roles" and prepare themselves to support and take over the tasks when their team mates become overloaded. An adaptive team, according to Kozlowski, is one that achieves "team efficacy", is confident to manage its own problems, and has the ability to change and adjust its structure (roles and tasks) and team processes "on-the-fly" to meet situational demands. The skills related to adaptive team performance are "contextually-based" and, therefore, must be developed in a work or performance context rather than through conventional training approaches like the classroom lectures. Kozlowski also emphasized "mastery learning goals" over "peformance-orientated goals". Mastery learning goals in Kozlowski's research focused on introducing trainees to different types of knowledge and skills. The learning episodes based on such goals are deliberately sequenced in a way that allow trainees to acquire the foundational knowledge first before learning the skills required for more complex situations. Kozlowski argued that sequenced learning goals facilitated self-monitoring and self-regulation and trainees are likely to achieve task self-efficacy. Performance oriented goals, on the other hand, focused on achieving specific levels of performance in sequenced training events. Kozlowski argued that performance goals inhibited learning and must be separated from learning goals. Trainees who were given mastery learning goals — compared to others who were given performance oriented goals — were found to pay more attention to developing situation awareness. In another study on leadership and enhancing team adaptability, Klein & Kozlowski (2008) reported that priming trainee's goals toward specific learning rather than performance objectives, and framing the training as an opportunity to build skills rather than as a demonstration of trainee's performance ability, improved their self-regulatory processes, learning and performance outcomes. Kozlowski recommended a framework of leader roles which will be discussed later in this chapter (see "Leader roles for developing adaptive expertise in teams").

Adaptive teams in the military. Klein & Pierce (2001) described another approach to understanding and developing adaptability in teams. They described adaptive teams as teams that "made modifications to meet new challenges" and as being "proactive and flexible". They suggested that teams adapt externally or internally. "External adaptation" is a process whereby adaptive teams use modular plans that they modify and change to meet the demands in the task environment. Teams require "outcome feedback" to check if their plans worked. "Internal adaptation" refers to a team's ability to restructure or reconfigure itself to meet the changing demands of the team or task environment. Klein & Pierce (2001) argued that internal adaptation is a form of "metacognition" where the team members needed to be able to observe themselves constantly to improve and adjust. For teams to internally adapt, they would need "process feedback". The authors proposed a number of skills and abilities that made a team adaptable. Examples of such skills and abilities include (a) anticipating problems areas in order to allocate attention and information-gathering resources", (b) adopting a mindset of expecting a plan to run into difficulties and to need modification", (c) "team competencies at routines that are building blocks of plans" and (d) having the experience in solving problems as a team. The authors also identified factors that could reduce adaptability in teams. Some of the practices or factors that reduce a team's capability to adapt include (a) the training of individuals and teams to master task routines, (b) the emphasis on the overlearning of skills, (c) assessing performance that was based on outcomes rather than processes, (d) centralizing so that there was control and (e) developing highly detailed and efficient plans. While such approaches and practices might have been effective in traditional military organizations that were structured to fight differently and respond in predictable ways in the past, they are no longer considered appropriate or valid to prepare military teams that need to respond dynamically in complex environments.

Critical Thinking Skills for Team Decision Making

Critical thinking is an individual level cognitive process. Team researchers believe that critical thinking in teams enhances the quality of team decisions and that critical thinking is especially important when decisions have to be made under complex and stressful conditions. It is important that team leaders understand what critical thinking does for teams in a decision making context and what they can do to initiate critical thinking process in their teams to enable better team decisions.

How are critical thinking skills developed in individuals and teams in the military? What does critical thinking involve and how does it help teams make better decisions in complex situations? Guillot (2004) proposed a definition of critical thinking for military professionals stating that critical thinking was "the ability to logically assess the quality of one's thinking and the thinking of others to consistently arrive at greater understanding and achieve wise judgments". Equipping team members with critical thinking skills will enable the team members to evaluate each others' assessments and the team as a whole to objectively evaluate all the factors in the task environment to achieve the best possible solution or decision for the team. Critical thinking skills function to prevent erroneous or faulty assessments and evaluations that lead to poor or bad decisions. Developing critical thinking skills has to be done in concert with the model of decision making adopted by the military (Cohen, Freeman & Thompson, 1998). It is, therefore, necessary to understand the basic ideas in decision theories which provide the context for a discussion on the role of critical thinking skills in team decision making.

Decision-Making Theories. Two key approaches dominate the literature on decision making. These approaches are generally referred to as the Classical Decision Making (CDM) theory and the Naturalistic Decision Making (NDM) theory. In conventional military operations, military planning and decision-making were deliberate processes of analyzing available facts or data or situation and devising detailed plans, which include planning for contingencies. This approach to military planning and decision-making adopts the CDM approach which had effectively guided military decision making in the past. The assumption underlying the CDM approach is that information and data are readily available, and are complete and clear such that decision makers can almost predict the outcomes of particular decision options. Such an approach will work well for military leaders if there is sufficient time to plan an operation. This is not the case in the 21st century where operations are usually carried out suddenly and under tremendous time pressure with highly complex or confusing information. The alternative approach to CDM is the NDM approach. In the NDM approach, the operating environment is described as being "dynamic and emergent, responsive to decision-maker actions, and subject to unpredictable shifts, resulting in significant time pressures and high costs for mistakes" (Cohen et al., 1998). The NDM approach does not require detailed planning on accurate and complete data but capitalizes on the experiences of decision makers. It is assumed that decision makers have the relevant experiences in the first place or work with others who have such experiences. In the last two decades, there has been an increased level of interest amongst military decision makers in NDM models. NDM models are based on a "pattern recognition" concept which refers to the use of cues or patterns that expert decision makers can recognize based on prior experiences in similar situations. Cohen et al. (1998) argued that while pattern recognition might serve decision makers in routine and similar situations or operational environments in the future, pattern recognition became useless when the cues or patterns were not present in novel task environments that decision makers encountered in complex environments.

To overcome the problems of previous NDM models, Cohen et al. (1998) proposed a Recognition/Metarecognition or R/M model. They argued that proficient decision-makers were firstly "recognitionally-skilled" and that secondly, they used processes that "verify and improve" the results of their recognition. They referred to these higher level processes as "meta-recognitional". According to Cohen et al. (1998), meta-recognitional processes help decision makers to:

- identify evidences for arguments based on the evolving situation
- identify problems in arguments supporting a conclusion or position taken about the situation which could surface "incompleteness", "conflict" and "unreliability" in the arguments
- take actions to correct arguments which may require additional data, taking alternative views and revising assumptions and
- to do a "quick test" that indicates when it is no longer appropriate to critique and correct; that a decision has to be made.

Cohen et al. (1998) suggested that the meta-recognitional processes were shaped by past experiences of success and failure and that experienced decision makers used such processes when the costs of delay were acceptable, when faced with uncertainty or when there was a high risk of errors. The researchers also argued that meta-recognitional processes controlled and even prevented responses based on recognition while they facilitated critical thinking processes to help decision makers to sieve out relevant information, take note of data that was contradictory, review their assumptions, reinterpret information and finally, to execute timely decisions.

Training critical thinking skills. According to Cohen et al. (1998), the R/M skills could be learnt to improve both decision-making processes and the outcomes of the decisions. The researchers developed training methods, tools and contents of a critical thinking training strategy based on the R/M model. They used a method known as Cognitive Task Analysis (CTA) which involved critical incident interviews where officers described their real operational experiences in making decisions

FIGURE 8-5. The STEP process

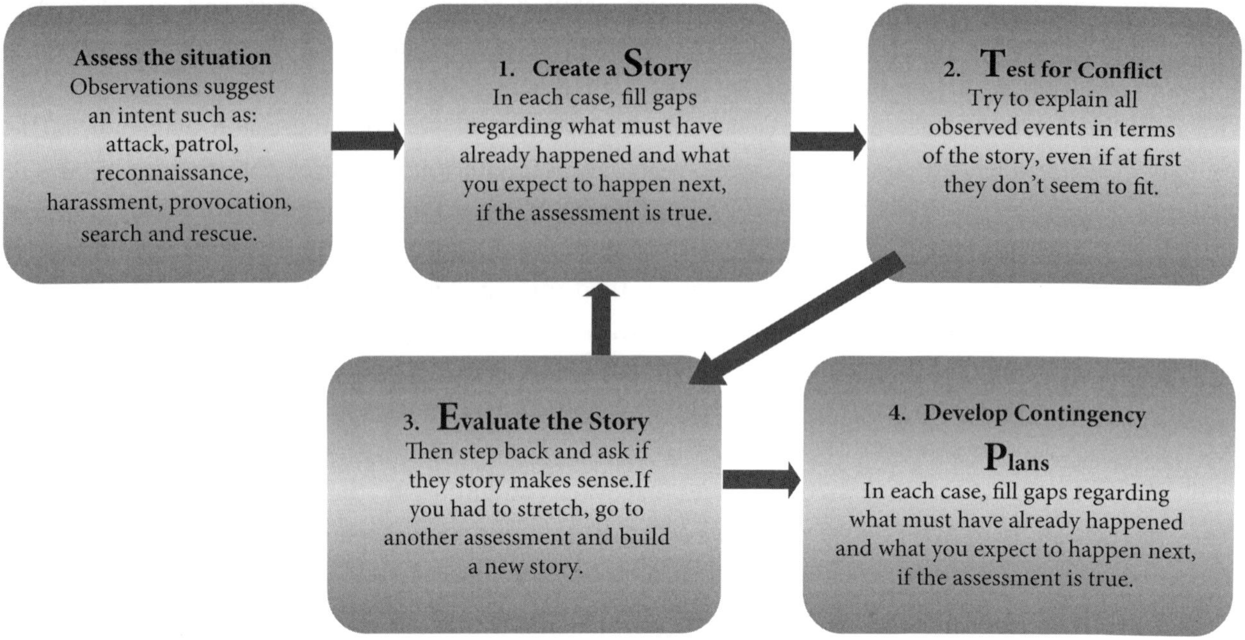

Source: Cohen et al. 1998

under conditions of uncertainty. The interviews were then analyzed to find out the decision makers' thinking strategies, how they organized information and the actual decisions that were made. Other methods used in developing the critical thinking training strategy included simulation, feedback, training guidelines and performance measures. A combination of lecture and discussions with guided practice and behavior modeling were adopted as methods of training. The content of the critical thinking training involved building stories in novel situations using a **Story-Test-Evaluate-Plan** or **STEP** process (see Figure 8-5 above), detecting and handling conflicting evidences through hostile intent stories, generating and evaluating alternative assessments by critiquing stories developed by the team and adjusting to available time. Results from Cohen et al. (1998) study showed that critical thinking training significantly enhanced decision making processes. The participants who received the training made more accurate decisions, generated more arguments to defend their assessments, identified more conflicting evidences and, generated more alternative assessments and contingency plans than participants who were not given such training. Those who received the training also produced assessments that were in agreement with the subject matter experts. Although the formal training of critical thinking skills is carried out at the individual level, there are two practical techniques that have been identified for team leaders to facilitate critical thinking in teams (see "Leader actions that develop critical thinking" in this chapter).

Shared Mental Models (SMMs)

In a Command Information Center (CIC) or an Air Traffic Control tower, members of the team can have different ideas of the task situation based on the information he/she may have on their data displays. Each of them can come to different conclusions about an unfolding reality and provide incorrect inputs that can lead to poor team decisions. It is important that team members have a shared interpretation of reality so that they can effectively coordinate the actions to deal with the situation. Such shared understandings in teams are referred to as shared mental models (SMMs). The concept of SMMs as a key team competency has consistently been associated with effective team performance under high-stress, high-demand task environments. Researchers found that SMMs played a key role in high performing teams and such teams tended to depend on SMMs particularly in high-stress, high demand situations. Cannon-Bowers, Salas & Converse (1993) suggested that SMMs helped teams to adapt quickly to changing task demands and enabled team members to predict what their teammates would do in terms of task performance and what they would need in order to do it. They argued that SMMs played a role in how team members coordinated their behaviors and what they selected as task strategies especially under situations that prevented team members from coordinating their tasks explicitly. These views were supported by other researchers who had observed effective teams sustain performance even under conditions of high workload, remaining effective even when there was little chance for team members to communicate openly with each other (Serfaty, Entin & Johnston, 1998). Under stress, when teams do not have the time to communicate openly, SMMs play a critical role by facilitating implicit communication and coordination processes, helping team members to foresee information and resources that their teammates would need and to respond appropriately.

Definition and types of SMMs. Mathieu, Heffner, Goodwin, Salas & Cannon-Bowers (2000) defined mental models as "organized knowledge structures that allow individuals to interact with their environment". According to Mathieu et al. (2000), mental models enable people to "predict and explain the behavior of the world around them", "recognize and remember relationships among components of the environment" and "construct expectations for what is likely to occur next". A team can have more than one SMM and these can include shared knowledge about (a) technology or equipment, (b) the team task and routines, (c) other team members' responsibilities, and (d) potential situations that the team may encounter (Cannon-Bowers et al., 1993). Cannon-Bowers & Salas (1998) have also suggested that highly effective teams held SMMs of situation and task environment and mutual mental models of interacting members' tasks and abilities that generate expectations about how other team members will behave. SMMs can be categorized into task-related or team- (process) related SMMs (Mathieu et al., 2000).

Developing SMMs in teams. The role and influence of SMMs in teams must be understood by team leaders so that they can actively develop and leverage on SMMs to prepare their teams for high-stress, high-demand situations. There is currently strong evidence to support the development of SMMs to enhance team effectiveness in 21st century military operations. Cannon-Bowers & Salas (1998) recommended that teams be given training to develop SMMs to enhance their shared knowledge of the team task and its demands, knowledge of the role relationships in the team and characteristics of team members. Training that deliberately shapes SMMs in teams will become a key strategy for adaptive team performance in 21st century military operations. It was also observed that leaders who communicated plans and strategies to their team members received more information relevant to the mission when under stress (Orasanu, 1990; Serfaty et al., 1998). The team training strategy that involves the development of SMMs is the Team Adaptation and Coordination Training or TACT strategy. In addition, leaders can also develop SMMs through specific leader actions which will be described later (see "Leader actions that develop SMMs in teams").

STRATEGIES FOR TRAINING ADAPTIVE TEAMS

Adaptability, critical thinking and SMMs are three of the competencies that teams need in order to respond effectively to operational challenges in complex environments. What are the training strategies that effectively develop these competencies of adaptability and SMMs in teams? To understand how these competencies are developed in teams, brief descriptions of the team training strategies and techniques developed and tested by TADMUS researchers are highlighted in this section. The role of team leaders in developing the competencies as well as the leader skills and behaviors that facilitate self-correction of team processes and support team learning will also be discussed. Readers interested in the details of the TADMUS research studies can refer to the edited book, *Making Decisions Under Stress: Implications for Individual and Team Training*. Figure 8-6 shows the five types of team training strategies that will be discussed.

FIGURE 8-6. Team training strategies.

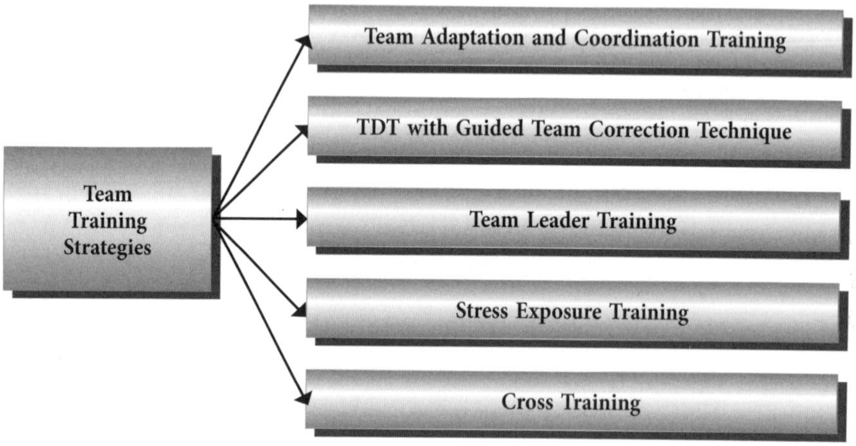

Team Adaptation and Coordination Training (TACT)

Teams do not naturally adapt and coordinate their tasks effectively under stress and uncertainty. They need to be trained and equipped with relevant team skills to enable them to adapt and coordinate under complex situations. TADMUS researchers, Serfaty et al., (1998) developed a team training approach called "Team Adaptation and Coordination Training" (TACT) to study the effectiveness of the strategy in enhancing adaptation and coordination processes in teams. This strategy also describes how adaptability and SMMs can be developed to enhance team performance under stress.

The TACT strategy is based on two theoretical models. The first is about the "adaptive team model" where it is suggested that a team adapts the way decisions are made, the way work is coordinated and reconfigures its structure in order to maintain performance under stressful conditions. In the TADMUS research, the team's perceived level of stress is managed at "tolerable levels" while the team attends to critical team processes to maintain performance. Serfaty et al. (1998) argued that adaptive teams "sensing strategies" and "related adaptation strategies" equipped the team to be sensitive to the stress. Based on this argument, the TACT strategy trained team members to sense and respond to the changes

Team training involves enhancing adaptation and coordination processes amongst members.

Source: CyberPioneer, Ministry of Defence, Singapore

in the task environment. The second model is about the role of SMMs in team coordination. Serfaty et al. (1998) proposed that effective teams had SMMs of the task environment and of the task and mutual mental model of other team members' tasks and abilities which facilitated team communication and coordination. In the TACT training, participants were given adaptive communication and coordination skills to foster teamwork and these included monitoring and back-up behaviors. They were also given a shared situation assessment (i.e. to develop their SMM) to facilitate implicit ways of communicating and coordinating. The researchers varied the stress levels in terms of high or low workload and ambiguity conditions. The results of the study showed that TACT training improved team performance under stress significantly. The researchers found that TACT training and, in particular, shared situation assessments facilitated implicit ways of communicating and coordinating, resulting in teams devoting more time to the operational task and wasting less time in using explicit modes of communication and coordination. Shared situation assessments proved to be an effective way to develop SMM in teams. The researchers also found that TACT training made teams more resilient to stress while also improving teamwork skills. The TADMUS study demonstrated that TACT training was an effective training strategy for developing adaptive responses in teams. Teams that received the TACT training were able to change their communication strategies and maintain effective performance under the stressful situations.

Team Dimensional Training (TDT) and Guided Team Self-Correction Technique

In complex military operations, team leaders do not have the opportunity to observe and correct team processes or to suggest where the team needs improvement (Smith-Jentsch, Zeisig, Acton & McPherson, 1998). Therefore, team leaders must enable their teams to objectively assess their own team processes, discover their own weaknesses, encourage and empower them to solve their own problems (Smith-Jentsch et al., 1998). A team's ability to self-manage, self-diagnose and self-correct team processes will be essential for its survival and performance in complex, 21st century environments. Team leaders must build such team capabilities before the team is considered operationally ready for complex military operations. Team Dimensional Training (TDT) with Guided Team Self-Correction Technique is a team training strategy developed by Smith-Jentsch, et. al., (1998). The researchers developed a framework and a process for leaders to help their teams achieve the internal capacity to self-manage team processes. The TDT (see Figure 8-7) was developed to enhance teamwork in Navy warships and accelerate their improvement on the four teamwork dimensions that relate to team performance. The TDT dimensions emphasize what teams must attend to in order to improve team processes in the next performance. The strategy requires team leaders to guide the team in self-critiquing and self-correcting team processes within a teamwork cycle consisting of four stages: (a) Prebrief, (b) Perform/Observe, (c) Diagnose Performance and (d) Debrief. In order to use this

FIGURE 8-7. Team Dimensional Training framework.

INFORMATION EXCHANGE	COMMUNICATION
Effective information exchange allows the team to develop and maintain shared situation awareness. **Components information exchange**: • Utilizing all available sources of information • Passing information to the appropriate persons without having to be asked • Providing periodic situation updates which summarize the big picture	While information exchange deals with what is passed to whom, this dimension involves how that information is delivered. **Components of information delivery**: • Proper phraseology • Completeness of standard reports • Brevity / Avoiding excess chatter • Clarity / Avoiding inaudible communications
SUPPORTING BEHAVIOR	**INITIATIVE / LEADERSHIP**
Supporting behavior involves compensating for one another in order to achieve team objectives. **Components of supporting behavior**: • Monitoring and correcting team errors • Providing and requesting backup or assistance to balance workload	Initiative / Leadership refers to anyone in the team who demonstrates initiative and leadership. **Components of initiative / leadership**: • Providing guidance or suggestions to team members • Stating clear and appropriate priorities

Source: Smith-Jentsch, Zeisig, Acton & McPherson (1998)

training strategy, team leaders are required to plan and design short exercises lasting not more than 30 minutes and include in these exercises team activities that require team coordination. Smith-Jentsch et al., (1998) argued that effective teams reflected on their past performance together and "self-corrected" after a critical event. Ineffective teams might have a number of reasons not to self-correct amongst which was the lack of diagnostic skills to provide feedback to one another constructively and a team climate that was not open for such feedback to take place. The researchers suggested that one way to help teams self-correct was by providing feedback skills training to enable teams to "self-correct" more effectively. Specific team leader actions and facilitation requirements that support this training strategy are discussed later (see "Leader Actions for Adaptive Team Performance" below).

Stress Exposure Training (SET)

Understanding how teams react to stress is perhaps the first step in preparing and training teams for 21st century challenges. If team leaders can prepare teams to deal with distractions in the task environment, cope with stress and focus their teams on their mission and tasks, then they would effectively have laid the groundwork for their teams to succeed in complex environments. Driskell & Johnston (1998) defined stress training as an "intervention to enhance familiarity with the criterion environment and teach the skills necessary to maintain effective task performance under stress conditions". Stress training provides knowledge about the stress environment, skills for maintaining effective performance under stress and builds performance confidence through practice under stress exposure.

How teams react to stress. Prior research has shown that an individual's breadth of attention can narrow as stress or arousal is increased and that an increased arousal can lead individuals to focus more on the self (Driskell, Salas & Johnston, 1999). The authors argued that team members who experienced stress might be distracted from the group task, adopted a "tunnel vision" and a more individual perspective of task activity. When a team is effectively performing under a stress-free condition, there are no problems for team members to take the larger team perspective. Under stress, however, team members will tend to become too focused on their own individual tasks, discarding the team perspective they might have held prior to the stress condition. In their study of three-person teams that performed a decision-making task under normal or high-stress conditions, Driskell & Johnston (1998) found that stress indeed narrowed the team perspective and team members under the stress condition were less likely to develop a strong team identity or adopt a team-level task perspective. Moreover, the results from the study also showed that a team perspective predicted task performance significantly. Driskell & Johnston (1998) concluded that under stress, the "collective representation of group activity" which the team members had, was crucial to interdependence in teams.

Stress Exposure Training (SET) strategy. Normal classroom type training conducted without stress conditions focused on training individuals or teams to master relevant skills without the interference of stress. However, individuals and teams are required to perform their tasks under real conditions of stress where they have to deal with distractions from unfamiliar situational factors. Stress training in the military prepares individuals and teams to identify stress symptoms, to manage stress levels and to ensure that performance under stress remains at the acceptable levels. Driskell & Johnston (1998) developed an integrated model of stress training known as SET which guides military training designers and leaders to consider important factors such as (a) when to introduce stress in training, (b) how to sequence the training, (c) when to introduce stress management skills and (d) what information related to stress should be given to the trainees. The SET framework provides knowledge of the stress environment, emphasizes skills training and instills confidence in trainee's ability to perform (Driskell & Johnston, 1998). The three phases of SET are described in Box 8-2 below. It is important for team leaders to understand the purpose of each phase and the contents of SET.

While the role of team leaders in SET has not been articulated, it is important that team leaders understand how teams react to stress and reinforce the teamwork skills under conditions of stress. Military team leaders could consider designing training that integrates some of the principles in the SET strategy.

BOX 8-2

Three phases of SET

Information Provision phase

There are two parts to this phase: (a) indoctrination and (b) preparatory information.

- "Indoctrination" provides information about the benefits of SET and the outcomes of ineffective or poor performance under stress, explains the importance of SET and gains trainees' motivation for SET.
- "Preparatory information" educates individuals about stress, stress symptoms and likely effects of stress in operational environments to increase the individual's "sense of controllability" as well the confidence in their ability to perform. The information develops accurate expectations about how trainees would react to stress so that they are less surprised or distracted by novel sensations, thus enabling them to task-relevant stimuli. In addition, preparatory information regarding the effects of stress on performance allows trainees to anticipate performance errors that are likely to occur in the stress environment. The three types of preparatory information are:
  - "sensory information" regarding physiological changes one is likely to experience (for example, increased heart rate, shallow breathing and muscle tension),
  - "procedural information" describing the events that will take place in the stress environment including likely stressors and their effects,
  - "instrumental information" providing instruction on what to do to mitigate or counter the stressful experiences to help the trainees manage the stress.

Skills Acquisition phase

In this phase, trainees are given behavioral and cognitive skills training to maintain effective performance in a stress environment. There are two categories of skills, (a) strategies that build resilience to effects of stress and (b) strategies that compensate for the losses in the task situation. The strategies include cognitive control strategies, physiological control strategies, overlearning, mental practice, time-sharing skills, decision-making skills and team skills. Team skills training include learning the importance of teamwork skills, how stress can affect teamwork and the practice of team skills, for example, how to give feedback to other team members. These teamwork skills are practiced in a simulated but realistic task environment.

Application and Practice phase

In this phase, training is designed to allow trainees to practice all the skills that were taught in Phase 2 under realistic task environments. Trainees practice the skills with increasing levels of stress and under increasingly realistic task conditions.

Cross Training

The earlier section of this chapter discussed how SMMs played a critical role in enabling teams to communicate and coordinate under stress. One way in which SMMs can be developed is through cross training, which refers to a team training strategy in which team members learn the duties of other teammates (Volpe, Cannon-Bowers, Salas, & Spector, 1996; cited in Blickensderfer, Cannon-Bowers & Salas, 1998). Through cross training, team members acquire the "basic knowledge necessary to perform the tasks, duties, or roles of the other team members and the overall team framework" (Blickensderfer et al., 1998). Cross training establishes a "common bond", shared experiences and feelings that are important for the "morale, cohesion, and confidence" in a team (Blickensderfer et al., 1998). The authors add that cross-training encourages members to "learn and understand the operations of those around them" thus enabling effective team behaviors such as "communication", "coordination" and "self-control". According Blickensderfer et al., (1998) cross training develops interpositional knowledge (IPK) which is "knowledge about the roles, responsibilities and requirements of other team members". IPK builds a clear understanding of how the team functions as a unit and how the individual team member's tasks and responsibilities interrelate with those of the other team members. According to the authors, cross training fosters the shared understanding and develops SMMs when team members acquire IPK in their team. The studies in the TADMUS program demonstrated that cross-training as a team training strategy is an effective way to develop IPK and that cross trained teams performed better in terms of teamwork, communication and task performance. Cross-trained team also maintained their performance better under stress. There are three types of cross training which vary in terms of the depth of information pertaining to the roles in a team and which should be selected based on the degree of interdependency in the team (Blickensderfer et al, 1998):

- **Positional Clarification** — training that raises awareness and provides team members with general knowledge of each other's role and associated responsibilities and what is required at the team-level. This type of training is conducted with discussions, lectures and demonstrations. Such training can be easily integrated into existing military team training programs.
- **Positional Modeling** — training which involves the discussion and observation of the duties of the other members where they learn more detailed information about how the team works, and how the duties interrelate and affect other team members. The training is designed around task work and trainees learn "concrete behaviors" that are required in task work processes. This training approach which involves a deeper understanding of team interdependencies is a useful approach especially for military teams that have a longer time frame to develop the deeper understanding of how the teams work.
- **Positional Rotation** — Training where trainees directly experience the team member's work through "hands-on practice". The training is aimed at getting members to a certain level of competency for basic tasks, including understanding the interactions and developing different perspectives of tasks

requiring high levels of cooperation and interdependency. The training must demonstrate the "team inter-dependencies" in order for team members to understand how they are interconnected to become the team (Blickensderfer et al., 1998). This approach will develop in teams not only the SMMs but also the confidence in team members to engage in back-up behaviors and to take over or cover the tasks of a disabled team mate. The approach enables the teams to maintain effective performances under high-stress and critical situations.

Although the research did not focus on the role of leader in cross training, the principles of cross training are useful, practical and important, for improving IPK which in turn facilitates the development of SMMs in teams.

LEADER ACTIONS FOR ADAPTIVE TEAM PERFORMANCE

Up to this point, three key team competencies and five approaches of team training featuring the training and development of the competencies to enhance adaptive team performance under stress were described. These cover the two big areas in terms of the content knowledge to lead teams in complex environments. The next section of the chapter synthesizes the knowledge in terms of concrete and practical actions which leaders can take to prepare and train their teams for complex environments.

Leader Behaviors that Support Team Learning

While the earlier sections provide leaders with the knowledge in terms of concepts, frameworks and techniques, this section is devoted to leader's own preparation for leading teams in complex environments. TADMUS researchers Tannenbaum, Smith-Jentsch & Behson (1998) argued that the effectiveness of 21st century military was a function of how well leaders were trained to support team learning. The earlier part of this chapter presented a diagram (see Figure 8-7) on the four knowledge and skills components that team leaders must have to lead teams in 21st century military operations. Developing leader behaviors and skills that support team learning is an area that will be high leverage for military organizations.

Tannenbaum et al. (1998) stated that the role of a team leader in facilitating team learning and the processes involved in team learning included (a) supplying information directly to the team, (b) establishing a conducive climate for learning, (c) providing support for "formal training experiences" and (d) facilitating or encouraging team discussions. The performance of these roles require leaders to adopt a facilitative approach with very different attitudes and behaviors from the usual "directive" or "command" approach that is typical of hierarchical, military organizations. In order to build confidence in adopting a facilitative approach for team learning in the military, leaders must be educated about the strengths of facilitation in a learning situation and how facilitation can strengthen rather than replace or weaken their command authority. In addition, military team leaders must understand that while the command or directive approach has its place in the military, adopting such an approach will likely destroy a team's motivation to learn.

Tannenbaum et al. (1998) proposed that leaders use a team learning cycle to support team learning. The four stages of the team learning cycle are (a) Clarification of Plans/Goals (pre-brief), (b) Performance/Practice, (c) Diagnosis of Performance and (d) Post-action review stages. Box 8-3 describes what leaders need to attend to in each stage so that team learning can occur. While it is important that leaders pay attention to what they need to do at each stage of the team learning cycle to support team learning, it is equally important for leaders to be aware of the following behaviors which, according to Tannenbaum et al. (1998), promote team learning. These behaviors include the leader:

- giving a self–critique at the beginning of the post-action review
- accepting feedback and suggestions from others
- avoiding person-oriented feedback
- focusing on task-focused feedback
- providing specific, constructive suggestions
- encouraging team members to actively participate during briefings and reviews and not simply share one's own observations and interpretations of the team's performance
- guiding the briefing session such that the team discusses both teamwork and task work processes
- referring to previous pre-briefs and team performances when facilitating subsequent debriefs, and
- letting the team know when they demonstrate good performance and show improvements.

Leader Roles that Develop Adaptive Expertise in Teams

Team leaders have an important role in developing team competencies and playing a role in each of the team training strategies that were described earlier. According to Kozlowski (1998), the adaptive team is a "self-learning system", where leaders were key to the developmental process. What leaders must do to develop adaptive expertise was articulated in a framework of leader roles which Kozlowski suggested were critical to the development of "adaptive expertise". The framework of four leader roles are:

- **Mentor.** As a mentor, the leader fosters the "development of shared affect", promotes the "understanding of the team's mission and goals", and develops bonds within new teams.
- **Instructor.** As an instructor, the leader attends to the acquisition of "task knowledge, proficiency, cognitive structures and self-efficacy" in novice teams.
- **Coach.** As a coach, the leader emphasizes team efficacy, develops SMMs and compatible behaviors that help the team to become more refined in their tasks as they become more mature and competent as a team.

BOX 8-3

Leader Actions in the Team Learning Cycle

(Tannenbaum et al., 1998)

Prebrief Stage

- Motivating team members to learn as a team.
- Introducing the TDT dimensions and behaviors to give the team a common language to discuss teamwork.
- Emphasize ***process-related*** rather than outcome-related goals.
- Creating a positive learning climate.
- Partner members in the learning process.
- Demonstrate genuine interest in team members' contributions and their development.

Performance Stage

- Observing team performance without interrupting the process.
- Noting mistakes and how they affect team outcomes.
- Adopting "assessment" rather than a "training" approach.
- Collecting examples of effective and ineffective behaviors across multiple team members.
- Collecting examples of behaviors related to success or failure in a team.
- Selecting examples to illustrate learning objectives in the TDT framework.

Diagnosis of Performance Stage

- Attending to individual members:
  - to discuss effective behaviors that helped the team,
  - to follow-up actions to improve team performance in future and,
  - to help each member set specific goals for improvement.
- Attending to the data to study the trends in the performance.
- Discussing the data with other leaders.
- Selecting examples of behaviors that are suitable for team discussions.
- Selecting one strength and a specific goal for each of the four dimensions.
- Helping team think of how to improve teamwork processes in subsequent exercises.
- Using opportunities to shape team mental models (or SMMs) of team processes.

Post-Action Review

- Reiterating objectives, setting stage and prepares team for discussion by outlining process, format and linking dimensions of the TDT to key events during the team performance.
- Displaying supportive leader behaviors to motivate team learning. Examples are:
  - motivating and encouraging team members to participate and ask questions
  - maintaining eye contact with the team members while facilitating the debrief
  - asking members to think of examples based on the TDT framework
  - restraining from giving any example before the team members do
  - giving more air time to team members as soon as they start contributing their points to the team discussion
  - allowing sufficient time for team members to think and respond before asking for their inputs
  - asking for positive examples first before negative ones
  - asking team members to identify corrective actions and waiting for someone in the team to provide an acceptable solution to the problem before moving on to the next topic

- **Facilitator.** As a facilitator, the leader helps the team to "self-manage" by giving assessments and providing feedback which expert teams need in order to maintain "team coherence", which according to Kozlowski is the foundation for team coordination and adaptability.

Adopting the appropriate roles to develop adaptive expertise in teams requires military leaders to first understand the developmental stages of their teams. For example, the team leader must know whether the team is at a novice stage before he/she could adopt the role of an instructor. The framework of the roles would, therefore, require that the team leader to closely monitor the team's development and level of competence. The ultimate objective of the leader is to help the team achieve "team coherence", where the team "self-manages", coordinates and adapts on its own to meet the demands in its operating environment.

An adaptive team has self-learning capabilities.
Source: Robert Nickelsberg/Getty Images News/Getty Images

Leader Actions that Develop Critical Thinking

Leaders can develop critical thinking in their teams by consistently practicing two techniques that were suggested by Cohen et al. (1998) to be important in developing critical thinking. These techniques were developed based on Cohen et al.'s (1998) critical thinking training strategy which was discussed earlier:

- **Situation Updates.** Team leaders in the research by Cohen et al. (1998) were trained to give "situation updates" which were designed to specifically highlight "missing, unreliable or conflicting evidence". The findings from the study indicated that the situational updates by the team leaders created a "shared mental model of the tactical situation" and developed a "shared metacognitive model of ongoing uncertainties". Cohen et al. (1998) tested the technique at the Surface Warfare Officers School (SWOS) and found that team members volunteered relevant information and insights without any prompting by the team leaders. The findings demonstrated that providing situational updates proved to be an effective training strategy for developing critical thinking skills in teams. Situational updates developed SMMs which strengthened decision making processes in a team.

- **Use of the Devil's Advocate Technique.** Team leaders could train and practice their team members to use the devil's advocate technique during a team decision making situation particularly to surface weaknesses in a team's arguments or assessments. Also referred to as the "crystal ball" technique, the technique involves a conclusion or argument made by the team and an appointed "devil" who needs to keep pointing out that the "conclusion is wrong" until no alternative explanation could be provided by the team members. The devil's role is to challenge team members to think of loopholes in their assessment and conclusions before final decisions are made based on the most credible story. Cohen et al. (1998) found that with this technique, team members were able to generate alternative assessments and interpretations of evidence. The training enabled team members to critique stories developed by the team.

Leader Actions that Develop SMM in Teams

By developing SMMs in teams, leaders can influence team decision making, and team adaptation and coordination. As earlier discussed, providing situational updates that highlight "missing, unreliable or conflicting evidence" helps teams think critically and develop an accurate SMM of the tactical situation. Leaders could also provide regular situation assessments or updates to help their teams maintain a shared situation assessment (or SMM) of a developing situation. The SMMs developed as a result of the leader's updates can affect how a team coordinates and changes its decision-making strategies under stress. Under stress and high workload teams can lose focus on their task and have a different understanding of the tactical environment. Leader briefings or periodic situation updates by team leaders help teams to "crystalize and confirm the common knowledge and situation mental model held by teams" (Serfaty et al., 1998). Research confirms that SMMs shaped by leader briefings positively influence communication processes and the performance of teams (Marks, Zaccaro & Mathieu, 2000). The research also highlights the importance of leader communications particularly in "sense-making contexts" where the leader plays a critical role in helping the team interpret and organize new information (Fleishman et al., 1991; Marks et al., 2000).

Leader Actions for Team Self-Diagnosis and Self-Correction

Team leaders play an important role in faciliating team self-diagnosis and self-correction of team processes. In the guided team self-correction process, leaders can help teams manage and train themselves, evaluate their team performance, identify the fundamental causes of problems related to team performance, develop solutions and plan future tasks. Leaders must keep the team focused on the team discussion, create a conducive climate, motivate and support active participation, role-model effective feedback skills and coach team members on how to put across their suggestions in a constructive way. By guiding the team to self-correct and self-manage, leaders also facilitate the development of accurate SMMs of teamwork among team members. Leaders will need to learn how to use the TDT framework, design short exercises targeted at surfacing team coordination behaviors and debrief teams using the data collected from observing team members during an exercise. In addition, team leaders manage team discussions during the debrief stage by (a) coaching team members to link behaviors to the components in TDT model, (b) allowing members to surface what needs to be corrected rather than impose his/her corrections, (c) communicating own interest in team's input, (d) facilitating members to address problems in teamwork and take part in critiquing their teamwork and (e) encouraging team members when they admit mistakes. Finally, team leaders need to help their teams identify their strengths and weaknesses, set goals for improvement and help the team revisit the goals in subsequent TDT pre-briefs.

SUMMARY

The key ideas to take away from this chapter are that military teams need to prepare for complex environments; that the preparation for 21st century military operations goes beyond team building and requires team training to embed new team competencies; that military leaders need to learn new knowledge, behaviours and skills in order to develop adaptive, learning teams.

Building on the foundational knowledge of leading teams based on group dynamics and OD traditions covered in Chapter 7, this chapter has introduced more advanced knowledge related to leading teams to perform adaptively in highly complex operating environments. This line of knowledge has a practical basis in concerns related to the complex challenges faced by teams in the aviation, medical and military contexts e.g., crew resource management and the U.S. Navy's tactical decision making under stress (TADMUS) program. The first section of the chapter highlighted three critical competencies needed in such teams that military leaders need to pay attention to. These include adaptability, critical thinking and shared mental models (SMMs). Adaptability in teams enables teams to deal with novel situations to be expected in complex environments. Critical thinking skills enable better situational assessments that lead to best possible team decisions under complex and high-stress situations. SMMs enable implicit communication and coordination, saving precious time that teams could use to focus on mission-critical tasks.

The second section of the chapter introduced some training, strategies and techniques that were developed and tested in the TADMUS program. The training strategies and techniques provide military leaders with the relevant knowledge and important ideas to train and prepare their teams for effective performance in complex environments. The strategies that were discussed include team adaptation and coordination training (TACT), Team Dimensional Training (TDT) with Guided Team Self-Correction Technique, Stress Exposure Training (SET) and Cross Training. The chapter concludes by suggesting five sets of actions that leaders could take to support the development of adaptive teams for complex environments. Firstly, leaders need to build adaptive expertise in teams by adopting facilitative and supportive roles rather than only task directive roles. Secondly, leaders need to develop critical thinking in teams and strengthen team decision processes by providing situational updates and training team members to use the Devil's advocate technique. Thirdly, leaders must build SMMs that underlie adaptive team performance under complex environment. SMMs are shaped and developed through consistent and regular leader briefings as was done to develop critical thinking. Fourthly, leaders must facilitate and guide their teams in self-diagnosis and self-correction of team processes. This is achieved when leaders create a climate conducive for team learning and keep the team focused on improving their next performance. Finally, leaders need to model behaviors that promote team learning and these involve specific behaviors e.g., behaviors related to giving and receiving feedback and facilitating team discussions around team and task work.

Key Concepts

- Adaptability in teams
- Cross Training
- Critical thinking skills
- Devil's Advocate Technique
- External Adaptation
- Guided Team Self-Correction Technique
- Implicit Communication
- Internal Adaptation
- Recognition/meta-recognition (R/M) model
- Shared Mental Models (SMMs)
- Stress Exposure Training (SET)
- Team Coordination and Adaptation Training (TACT)
- Team Dimensional Training (TDT)
- Team learning
- Team training

Questions

- What are your experiences of team leadership and training teams in a technologically complex environment? How would you explain some of these experiences using the concepts and theories discussed in this chapter?
- How has your training as a military leader prepared you for team leadership in complex environments? What concepts and strategies have you learnt that could be applied in training and preparing your teams for complex, 21st century military operations?
- Reflect on your own knowledge, behaviors and skills to lead teams in complex environments. What knowledge, behaviors or skills discussed in this chapter did you find particularly interesting and useful for you to (a) help your teams self-manage and self-correct and (b) support team learning?

References

Blickensderfer, E., Cannon-Bowers, J.A. & Salas, E. (1998). Cross-Training and Team Performance. In J.A. Cannon-Bowers & E. Salas (Eds.), *Making Decisions Under Stress: Implications for Individual and Team Training* (pp. 299–311). Washington: American Psychological Association.

Canadian Defense Academy (2005). *Leadership in the Canadian Forces Doctrine.* Canadian Forces Leadership Institute, Canadian Defense Academy.

Cannon-Bowers, J.A., Salas, E., & Converse, S.A. (1993). Shared mental models in expert team decision making. In N.J. Castellan, Jr. (Ed.), *Individual and group decisions making: Current issue* (pp. 221–246). Hillsdale, NJ: Erlbaum.

Cannon-Bowers, J.A., & Salas, E. (1998). Individual and Team Decision Making. In J.A. Cannon-Bowers & E. Salas (Eds.), *Making Decisions Under Stress: Implications for Individual and Team Training* (pp. 17–38). Washington: American Psychological Association.

Cohen, M.S., Freeman, J.T., & Thompson, B. (1998). "Critical Thinking Skills in Tactical Decision Making: A Model and a Training Strategy". In J.A. Cannon-Bowers & E. Salas (Eds.), *Making Decisions Under Stress: Implications for Individual and Team Training* (pp. 155–189). Washington: American Psychological Association.

Driskell, J.E., & Johnston, J.H. (1998). Stress Exposure Training. In J.A. Cannon-Bowers & E. Salas (Eds.), *Making Decisions Under Stress: Implications for Individual and Team Training* (pp. 191–217). Washington: American Psychological Association.

Driskell J.E., Salas E., & Johnston J.H. (1999). Does stress lead to a loss of team perspective? *Group Dynamics Theory Research & Practice*, 9(3), 291–302.

Fleishman, E.A., Mumford, M.D., Zaccaro, S.J., Levin, K.Y., Korotkin, A.L., & Hein, M.B. (1991). Taxonomic efforts in the description of leader behavior: A synthesis and functional interpretation. *Leadership Quarterly*, *2*, 245–287.

Guillot, W.M. (2004). Critical thinking for the military professional. *Air & Space Power Journal.* Retrieved from: http://www.airpower.au.af.mil/airchronicles/cc/guillot.html.

U.S. Army Research Institute. (2008). *Leadership: Enhancing Team Adaptability in Dynamic Settings* (Technical Report 1228). Washington, DC: Army Research Institute.

Klein, G., & Pierce, L. (2001). *Adaptive Teams.* Proceedings of the 6th ICCRTS Collaboration in the Information Age Track 4: C2 Decision-Making and Cognitive Analysis.

Kozlowski, S.W.J. (1998). "Developing Adaptive Teams". In J.A. Cannon-Bowers & E. Salas (Eds.), *Making Decisions Under Stress: Implications for Individual and Team Training* (pp. 115–153). Washington: American Psychological Association.

Kozlowski, S.W.J., & Ilgen, D.R. (2006). Enhancing the Effectiveness of Work Groups and Teams. *Psychological Science in the Public Interest*, 7, 77–124.

Lauber, J. (1984). Resource management in the cockpit. *Air Line Pilot, 53*, 20–23.

Marks, M.A., Zaccaro, S.J., & Mathieu, J.E., (2000). Performance Implications of Leader Briefings and Team-Interaction Training for Team Adaptation to Novel Environments. *Journal of Applied Psychology, 85(6)*, 971–986.

Mathieu, J.E., Goodwin, G.F., Heffner, T.S., Salas, E. & Cannon-Bowers, J.A. (2000). The Influence of Shared Mental Models on Team Process and Performance. *Journal of Applied Psychology, 85(2)*, 273–283.

Orasanu, J.M. (1990). Shared mental models and crew decision making, *CSL Report 46.* Princeton, NJ: Cognitive Science Laboratory, Princeton University.

Salas, E., Wilson, K.A. Burke, C.S., Wightman, D.C., & Howse, W.R. (2006). Crew Resource Management Training: Research, Practice, and Lessons Learned. In Robert C. Williges (Ed.), *Reviews of Human Factors and Ergonomics, Vol. 2.* Santa Monica, CA: Human Factors and Ergonomics Society.

Serfaty, D., Entin, E. E., & Johnston, J. H. (1998). "Team Coordination Training". In J. A. Cannon-Bowers & E. Salas (Eds.), *Making Decisions Under Stress: Implications for Individual and Team Training* (pp. 221–245). Washington: American Psychological Association.

Smith-Jentsch, K.A., Zeisig, R.L., Acton, B., & McPherson, J.A. (1998), "Team Dimensional Training: A Strategy for Guided Team Self-Correction". In J.A. Cannon-Bowers & E. Salas (Eds.), *Making Decisions Under Stress: Implications for Individual and Team Training* (pp. 271–297). Washington: American Psychological Association.

Tannenbaum, S.I., Smith-Jentsch, K.A., & Behson, S.J (1998). Training Team Leaders to Facilitate Team Learning and Performance. In J.A. Cannon-Bowers & E. Salas (Eds.), *Making Decisions Under Stress: Implications for Individual and Team Training* (pp. 247–270). Washington: American Psychological Association.

Van Der Kleij, R., & Schraagen, J.M. (2006). Enabling Team Decision Making. In C. Bowers, E. Salas & F. Jentsch (Eds.), *Creating High-Tech Teams* (pp. 39). Washington DC: American Psychological Association.

Volpe, C.E. Cannon-Bowers, J.A., Salas, E., & Spector, P. (1996). The Impact of cross training on team functioning. *Human Factors, 38*, 87–100.

9

Leading in Hierarchical, Learning Military Organizations

Source: Martin Bureau/AFP/Getty Images

OVERVIEW

Military Leadership occurs in context of an organization which is characterized by a unique structure and culture. A review of military leadership by Wong, Bliese and McGurk (2003) described the U.S. military at the start of the 21st century as "a traditionally hierarchical institution that finds itself in an uncertain, volatile world executing missions with very high consequences" (p. 660). Faced with highly dynamic security environments and rapid advances in technology, most military forces in the new millennium are trying to re-configure themselves as "learning organizations" that exist in a constant state of learning, change and transformation. Such transformational change is seen as an ongoing process rather than occasional — where change is treated as a constant "journey" rather than a "destination" or "end-state". What is the nature of leadership at different levels of a hierarchical organization like the military? How is the structure of military organizations changing and how can this affect military leadership? What are the challenges of leading in hierarchical yet continuously transforming, learning military organizations? This chapter aims to provide students with a basic appreciation of the nature and challenges of leadership beyond the face-to-face, "direct", or "team"-levels which were the focus of Chapters 5 to 8. To the extent that military leadership takes place in organizations that are traditionally hierarchical in structure, the chapter begins by reviewing the literature on "levels of leadership" — both in the academic literature and in military doctrines. The concepts of "organizational structure" and "organizational culture" are also introduced to help readers to appreciate organizational level concepts and thinking. Finally, the idea of the "learning military organization" is discussed, including the challenges and the role of leaders in operationalizing the idea of the "learning organization" in the military.

UNDERSTANDING ORGANIZATIONS

Organizational Structure

In the textbook entitled *Organizational Theory*, Robbins (1987) defines an organization as "a consciously coordinated social entity with a relatively identifiable boundary that functions on a relatively continuous basis to achieve a common goal or set of goals" (p. 3). All organizations are characterized by a certain "structure . . . (which) stipulates how tasks are allocated, who reports to whom, and the formal coordinating mechanisms and interaction patterns that will be followed . . ." (p. 4). Robbins (1987) adds that organizational structures differ in three main ways:

- **Complexity**: The degree of "differentiation" or specialization in the organization including the extent of division of labor and responsibilities, hierarchy, etc.
- **Formalization**: The extent to which organizational rules, practices are codified or documented in the form of a constitution, orders, directives, guidelines, etc.
- **Centralization**: The extent to which decision making is held by a few members.

The organizational theorist Gareth Morgan (1986) in *Images of Organization* suggests that the industrial age metaphor of "organizations as machines" was influenced by the way that Frederick the Great modernized his military forces in the late 1700s by introducing ranks and uniforms, increasing task specialization, and standardizing regulations, equipment and training. He wrote: "Frederick's aim was to shape the army into an efficient mechanism operating through means of standardized parts" (p. 16). By the 1900s, the modern sciences of management and organizations emerged with the publication of Frederick Taylor's (1911) classic *Principles of Scientific Management*, which advocated the scientific measurement of work and the standardization of tasks, the systematic selection, training and development of workers, and, a "division of labor" between management and workers in organization. In line with these developments, the U.S. Army also led the development and implementation of the first ever mass-administered IQ tests for the purpose of systematic personnel selection during the First World War (see Yerkes, 1921). In 1916, the French engineer Henri Fayol also published a book entitled *Administrative Science* which proposed 14 principles that all organizations should apply to run effectively (see Table 9-1). Many of these principles continue to be found in the "principles of war" of various military doctrines today. For example: "unity of command" in U.S. military, "unity of direction" which corresponds to "objective" in U.S. military, and "selection and maintenance of aim" in British forces. From this list of principles, it is not difficult to see how most military forces at the turn of the new millennium are still essentially "industrial-age" organizations, designed to function efficiently like machines. As Wong et al., (2003) suggested, "In terms of organizational form, the military is unquestionably traditional . . . (with a) clear delineation of power across hierarchical levels and clear prescriptions about how leaders and subordinates are expected to interact" (p. 659–660).

TABLE 9-1. Henri Fayol's 14 Principles of Organization

• Division of work (or Specialization)	• Centralization (of decision making)
• Unity of direction (common aim/ objective)	• Order (People and materials at right time & place)
• Discipline (of employees)	• Equity (fairness to subordinates)
• Unity of command ("one boss")	• Stability (of tenure)
• Authority (of the management)	• Initiative (originate ideas and show initiative)
• Subordination (of individual to organization's interests)	• Scalar chain (line of authority and communications)
• Remuneration (fair compensation)	• Esprit de corps (for harmony and unity)

Source: Fayol (1916).

Organizational Culture

Besides organizational structure which attempts to integrate organizational functions and activities to achieve efficiency, organizations also possess "cultures", which refers to the norms of behavior, the man-made artifacts including symbols, rules and routines, and the beliefs and values that emerge in the organization over time. Schein (1985) provided a well-accepted definition of organizational culture as "a pattern of basic assumptions — invented, discovered, or developed by a given group as it learns to cope with its problems of external adaptation and internal integration — that has worked well enough to be considered valid, and, therefore, to be taught to new members as the correct way to perceive, think and feel in relation to those problems." In short, some think of organizational culture as "the way we do things around here" (p. 277, Burke & Litwin, 1989). The elements of organizational culture include: values (both stated or "espoused" and unstated or "in-use"), expectations (including the roles and rules of behavior), the customs and rituals or "traditions", the stories, myths, metaphors and symbols concerning the history of the organization, the key figures, and the everyday shop talk or "lingo" within the organization. The terms "organizational culture" and "organizational climate" refer to different things — Schneider (2000) clarified that while organizational climate referred to the shared perception of the environment or setting in which people worked or the mood or "feel" of the workplace or organization, organizational culture concerned the attributions made about *why* the work environment or setting was the way it was, or the "thinking" or "logic" of the workplace.

Research reported in the business bestsellers *Good to Great* (Collins, 2001) and *Built to Last* (Collins & Porras, 1996) demonstrated that organizational culture could explain why certain organizations over time excelled within their industry, relative to their competitors. Together, organizational structure and culture provide a framework to guide individuals, teams and units within an organization to act in a consistent manner to achieve organizational goals. While both organizational structure and organizational culture have overlapping functions, each is not a substitute for the other. Indeed, both are needed in organizations.

The Nature of Leadership in Organizations

What is the relationship between leadership and organizations? One approach articulated by Schein (1985) in his classic book, *Organizational Culture and Leadership* was that leadership is one of the key factors that affected organizational culture. Schein (1985) argued that organizational leaders could shape culture both directly and indirectly. Direct ways include the issues or things that the leaders pay attention to in the organization, how the leaders respond to organizational crises and challenges, how they role-model their policies and beliefs, and, their criteria or considerations for selecting, recognizing and dismissing personnel in the organization. Secondary or indirect ways in which leaders can influence organizational culture include how they design the organization's structure including the lines of communication or reporting, the procedures or systems of managing resources and facilities, the stories, legends, and myths contained in their communications, and the formal statements that they make about the organization.

During the early 1990s, the U.S. Army Research Institute recognized that the most empirical research tended to study leadership in terms of a direct- or interpersonal-level social exchange in the absence of the organizational context. It therefore funded several research programs to gain a better understanding of the nature of leadership in organizations. This generated several key publications including conceptual and empirical reviews of "executive leadership" by Zaccaro (1996, 2001), and an enhanced theoretical understanding of leadership *in context* of organizations entitled *Nature of Organizational Leadership* (Zaccaro & Klimoski, 2001). In his initial literature review, Zaccaro (1996) concluded that all organizational leaders engage in three important classes of activities:

- **Direction setting**: This includes behaviors like envisioning the future, planning, strategizing and setting goals for other members of the organization.
- **Boundary-spanning**: This refers to behaviors that involve connecting individual members or groups or units outside of the organization.
- **Operational management**: This includes the routine performance-related actions like executing, maintaining and coordinating tasks within the organization.

Organizational leadership gets more complex at higher levels.
Source: Carlo Allegri/Getty Images News/Getty Images

Zaccaro highlighted that at higher levels of organizational leadership, direction setting would incorporate a longer time frame while boundary spanning would increasingly occur in the organization's external environment. At the highest, executive levels, Zaccaro argued that leaders needed to possess three important capabilities:

- **Flexible, cognitive integrative complexity**: This included the ability to consider multiple perspectives to an issue, and to establish connections across these different perspectives in a flexible (rather than a rigid) way.
- **Social behavioral flexibility**: This included the ability to perceive the subtleties of social situations, and behavioral flexibility, which included possessing a wide range of behaviors to respond to different social situations.
- **Dispositional orientation toward flexibility**: This included traits such as openness, curiosity and self-discipline.

Building on the above ideas, Zaccaro and Klimoski (2001) identified an expanded list of the challenges of organizational leadership, especially at the higher levels, including the following:

- **Cognitive challenges.** Organizational leaders need to deal with highly complex information processing and problem-solving demands, which are often the result from many different but related factors that affect the organization's success.
- **Social challenges.** Organizational leaders need a repertoire of social skills and abilities to deal with the wide range of social relationships in the leader's operating environment. This includes coordinating and supervising the activities of multiple individuals and units across the organization.
- **Personal challenges.** Besides focusing on the organization's mission and "bottomline", organizational leaders must also manage their own careers and reputation in a skillful and timely manner — especially important during periods of leadership or job transition.
- **Political challenges.** Organizational leaders must deal with the many political pressures that relate to power in organizations. This includes acquiring power and authority and employing these in a timely and judicious manner. Leaders in organizations must also routinely build and maintain coalitions; and they may have to deal with interpersonal or intergroup conflicts and dynamics, which — especially at top organizational levels — may critically affect organizational success.
- **Technological challenges.** To ensure organizational relevance and effectiveness, organizational leaders need to constantly keep up with technological advances which can result in fundamental changes in the organization's core business and functioning.
- **Financial challenges.** Ultimately, all organizations are run on budgets that need to be well-managed. Organizational leaders must deal with financial challenges from suppliers, customers and investors, workers and managers in a manner that is able to balance short term concerns with long-term, strategic objectives.
- **Staffing Challenges.** Organizational leaders must constantly prepare the future leaders of the organization, while also coordinating the contributions of multiple executives. They are also responsible for staffing organizational positions with individuals with the right set of skills, dispositions, and capabilities. Organizational leaders must therefore understand and influence the system of recruitment, selection, training, development, and performance measurement and management.

LEVELS OF MILITARY LEADERSHIP

As stated previously, Wong, et al., (2003) noted that a core characteristic of the modern military organization was that it was structured in a traditionally hierarchical form for maximum efficiency and stability. Given the dominance of hierarchies in most military organizations, it is useful to examine how military forces think about the nature of leadership at different levels of the organization. Such knowledge can serve to guide the systematic leadership development of military leaders as they advance in their seniority in the organization.

"Levels" in U.S. Army Doctrine

If one examines the evolution of the U.S. Army's leadership doctrine, one will note that until 1990, the U.S. Army had three different doctrinal documents on leadership: Field Manual 22-100 entitled *Military Leadership* (1983) which explicitly aimed to provide doctrinal guidance for its company grade officers and non-commissioned officers; Field Manual 22-103 entitled *Leadership and Command at Senior Levels* (1987) which was targeted at "leaders who must use both direct and indirect influence to achieve organizational success", and Department of the Army Pamphlet 600-80, entitled *Executive Leadership* (1987) which provided guidance for civilian and military leaders who "operate at the national and strategic levels". In 1999, the U.S. Army published a new edition of the Field Manual 22-100 which articulated three conceptual "levels of leadership". It explained that there are multiple factors which determine a leader's "leadership level". These include the position's span of control, the headquarters level, the extent of influence, the size of the unit or organization, the type of operations that the unit conducts, the number of people assigned, and the planning horizon (over time). These three conceptual "levels" remain till today in the U.S. Army's leadership doctrine, Field Manual 6-22. They are:

- **Direct-level leadership**. This is the level of leadership which is "close enough to the action" to be able to determine and address problems immediately; where the subordinates are accustomed to seeing their leaders most of the time; and, where the leader is able to exert direct influence, either by visiting the workplace or by interacting with subordinates during functions. It is also known as "face-to-face" or "first-line" leadership. For the U.S. Army, the span of influence of a direct leader may range from a few to a hundred personnel and their

Direct-level leadership involves face-to-face, first-line influence of people.
Source: CyberPioneer, Ministry of Defence, Singapore

planning horizon is short — typically from a few weeks to a year or at most two years ahead. The leaders of military teams, squads or sections, platoons, companies and battalions are mainly direct-level leaders. The actions of direct leaders include monitoring and coordinating team efforts, providing clear and concise instructions, setting performance goals and expectations, developing subordinates individually.

- **Organizational-level leadership**. This is the level where military leaders mainly influence people in their organization indirectly by establishing and integrating or aligning policies and systems, as opposed to direct, face-to-face contact. The planning horizon of organizational leaders generally ranges from two to 10 years, and their leadership influences hundreds to a few thousand personnel. They rely on their subordinate leaders to help them to motivate and inspire their people directly and to manage the organization's resources. This reliance on intermediate levels of subordinate leaders makes it more difficult for organizational leaders to ascertain or assess the results of their decisions immediately. On a routine basis, organizational leaders need to establish policies, manage multiple priorities and a wide range of resources. They must also establish a long-term vision for the organization and empower others to perform their missions. To achieve this, they need to establish an organizational climate to support subordinate leaders in their efforts to influence their subordinates. Because organizational-level leaders are physically distant from frontline operations, they must make special efforts to "get out of the office" to visit personnel in remote parts of the organization in order to understand the conditions that their people face and to verify if their policies and systems have the intended effects or outcomes in the "field". With such "field visits", organizational-level leaders may then determine if there is a need to re-evaluate their policies and the organization's priorities. On the whole, organizational-level leaders deal with more people, more complexity, more uncertainty, and more unintended consequences than direct level leaders.
- **Strategic-level leadership.** Strategic-level leaders are those who are responsible for large organizations, and who influence several thousands to even hundreds of thousands of people. Strategic-level leaders work in highly uncertain and complex environments. Their main tasks including establishing the organization's vision, strategy and structure, allocating resources, communicating the organization's vision, and preparing the organization for the future. To do this, they must process information quickly and assess alternative options, often with incomplete information, make decisions, and build leadership and organization-wide support for their strategies. The actions and decisions of strategic leaders may have global and sometimes even historical impact. The strategic level military often has to explain to the country's political leaders the military's missions, strategies and budgets. Long-term organizational change and transformation are the main concerns of strategic-level leaders. They have look into the development and acquisition of new military capabilities, and to possess vision for the future of the organization, yet, most of them never get to see the realization of their ideas or vision during their tenure. Strategic-level military leaders must also constantly balance a focus on change with a focus on current operations. They must be highly skilled in selecting and developing the future generations of leaders for the organization.

"Levels" based on Stratified Systems Theory

The above conceptual levels of leadership adopted by the U.S. Army are based on Stratified Systems Theory which was developed by Jacobs and Jacques (1987, 1990, 1991), and which was based on a broader, research-based theory developed by Jacques et al. (1978; see also Jacques, 1989).

Since the 1960s, Jacques had sought to determine a set of principles that could help management to design the "right" organization based on the nature of the work that it must perform. Jacques tried, for example, to establish a basis for determining just how many layers of hierarchy an organization should have for optimal efficiency and worker satisfaction. He interviewed and observed many different types of workers at different levels of organization and discovered that one could objectively determine the "size" of any organizational job or role based on the timespan of the longest task. More importantly, Jacques discovered that the timespan of tasks or responsibilities increased with the level of leadership or management in any organization. At the lowest levels, the longest tasks could be completed within days or weeks. At higher, "strategic" levels the longest tasks or responsibilities may not be completed for 20 years or more. Using timespan as a measure of work, Jacques proposed that one could define up to seven meaningful levels in organizations with each level providing a different context or set of demands on the leaders or managers at that level. Table 9-2 shows how the U.S. Army's three levels of leadership is related to Jacques' seven. Readers may refer to a technical report by Jacques and Stamp (1995) for more in-depth discussions on the application of Stratified Systems Theory in military organizations.

TABLE 9-2. U.S. Army's three levels of leadership with Jacques' seven levels.

Jacques' Level	Timespan	Main leadership tasks at this level	U.S. Army's levels
7	20+ years	Create complex systems; establish policy; plan and organize the acquisition of major resources.	Strategic leadership
6	10+ years	Oversee the operation of subordinate systems, apply organizational policy.	
5	5+ years	Directly operate complex systems.	Organizational-level leadership
4	2+ years	Plan and allocate resources to interdependent subordinate programs or units.	
3	1+ years	Develop and execute plans to implement policy or assigned missions.	Direct-level leadership
2	3+ months	Perform the job, including anticipating and solving problems in real-time.	
1	<3 months	Employ skills, practical experience and judgment to perform routine, hands-on work.	

Source: Jacobs & Jaques (1991); Jacques & Stamp (1995).

"Levels" in the Canadian Forces

Today, the Canadian Forces also makes explicit reference to Stratified Systems Theory as a basis for its leadership doctrine which distinguishes two levels of leadership: "Leading People" and "Leading the Institution" (Canadian National Defense, 2005a, 2005b, 2007a, 2007b). These two levels are essentially distinguished in terms of the ideas of direct versus indirect leadership. The Canadian Forces doctrine states: "leadership at the tactical level is primarily about accomplishing missions and tasks through direct influence on others — that is, leading people", while "institutional leadership, on the other hand, takes place predominantly in higher headquarters at the operational, or theater, level and at the strategic level of the CF. There is a predominant focus on indirect influence on organizational and campaign performance" (p. vii). The Canadian Forces emphasizes the *conceptual* nature of these two levels of leadership by de-linking these levels from rank. It stresses that both its senior officers and senior non-commissioned members are increasingly expected to lead at the institutional level. The Canadian Forces (2005b) also introduces the concept of "distributed leadership" as a way to distinguish leadership levels from (structurally- or hierarchically-defined) military rank by emphasizing a "shared" as opposed to a "heroic" view of leadership. It explains that "no one person can master and control everything in a large and complex organization. ... over the long haul, the effectiveness of the institution will depend more on developing in breadth and depth a strong Officer-NCM leadership team with a shared sense of responsibility, professional identity, values, and purpose" (p. 6). With this, the Canadian Forces (2005b) defines distributed leadership as the "sharing the responsibilities of leadership vertically and horizontally within teams, units and formations and within the CF as a whole" (p. 6). The concept of distributed leadership requires that all of its personnel think of leadership as a shared function or responsibility in their team or organization; that all personnel are expected to influence each other to achieve outcomes depending on the situation that they are in.

DO NEW FORMS OF MILITARY ORGANIZATION NEED MORE OR LESS LEADERSHIP?

In the late 1980s, organizational scientists started to notice changes in organizational forms and structure that were to a large extent enabled by rapid developments in information technology. In 1995, a group of organizational consultants published a book entitled, *The Boundaryless Organization: Breaking the Chains of Organizational Structure* (Ashkenas, Ulrich, Jick & Kerr, 1995, 2001) where they described the efforts of General Electric to transform itself from a rigidly hierarchical organization to a "boundaryless one". They argued that traditional organizations were characterized by four kinds of boundaries including clear divisions between different levels of management or "vertical" boundaries, clear lines between different organizational functions and divisions or "horizontal" boundaries, clear "external" boundaries between the organization and its environment (including customers, suppliers, etc.), and geographic boundaries between locations, regions and cultures. In their view, boundaryless organizations were those that had a culture that sought to overcome the traditional boundaries that tended to stifle the flow of knowledge, ideas and information among organizational members and with external agents. They argued that whereas traditional organizations tried to achieve success through factors like size, role clarity, specialization and control, newer organizations would define success in terms of factors like speed, flexibility, innovation and integration.

In the mid-1990s, the U.S. Army brought together many thought leaders in the study of organizational leadership where papers were presented on the nature of leadership in 21st century military forces (see Hunt, Dodge & Wong, 1999). Two of the participants, Shamir and Ben-Ari (1999), discussed three different scenarios often portrayed of military leadership in the 21st century, "digitized" battlefield. These included:

- **Disposable leadership**: In this scenario, the idea is that leadership will itself become "ad-hoc" or "disposable" as organizations rely more and more on temporary or ad-hoc arrangements for highly knowledge-based projects, often involving highly specialized personnel to "lead" others at different stages of the project.
- **Collective leadership**: The idea in this scenario is that leadership is no longer concentrated in a single person or group but instead shared or performed by many or even all members in the team — either simultaneously or sequentially.
- **Tele-leadership**: This scenario presents the military commander as standing in the middle of a complex information center; where the leader's role mainly involves reading and interpreting information conveyed primarily by electronic means. In this model, the leader influences others mainly via electronically transmitted orders and instructions with minimal human-to-human interaction.

Shamir and Ben-Ari (1999) noted a "blurring of the frontline" in 21st century military operations. They also postulated that military forces would have to work with other agencies in arrangements that were characterized by "temporary structures, fuzzy boundaries, and unclear divisions of labor and authority" (p. 24). In a separate paper addressing the question whether leadership would be more or less important in boundaryless organizations, Shamir (1999) argued that "far from becoming disposable, leadership becomes indispensable in the boundaryless organization" (p. 50). In this scenario, Shamir argued that leaders would become "'centers of gravity' in the midst of weakening frameworks, and balancing the centrifugal forces exerted by loosely coupled structures, fragmented cultures, temporary membership, and technologies that increase the distance between leaders and members". In his view, leaders of such "looser" organizations have to use their influence to play an "integrative" role by bringing the many human and technical elements to work together to achieve the organization's desired outcomes. Leaders may also have to resort to more indirect means such as shared mental models, values and frameworks to better achieve coordinated and collective action throughout the boundaryless organization. Shamir (1999) suggested five specific issues that leaders would need to pay attention to when leading in such non-traditional organizations. These include:

- Replacing or reinforcing the weakened organizational structures and cultures with frameworks or "mental models" so as to better guide and coordinate collective action.
- Increasing the level of commitment among organizational members especially when they are faced with constantly changing work which challenges their commitment.
- Creating and supporting cohesive work teams in the face of increased diversity and changing team membership as people transit more frequently in their careers.
- Providing all in the organization with a sense of "psychological safety" in order to help people to cope with the anxieties related to uncertainty and change.
- Supporting a climate that is conducive for individual and organizational learning.

Does "tele-leadership" mean less influence?
Source: Getty Images

LEARNING MILITARY ORGANIZATIONS

Soon after the September 11th terrorist attacks in New York in 2001, the U.S. government announced its "War on Terror" and made transformational changes in the way its forces were organized — both structurally and culturally. Today, various military forces make reference to the concept of a learning organization. In a document entitled *Canadian Officership in the 21st century*, also called "Officership 2020", the Canadian National Defense directed its military forces to "create the policies and conditions that would transform the Canadian Forces into a true learning organization". The vision was for the organization to be "skilled at acquiring and transferring knowledge and at encouraging behavior that reflects new knowledge and insights" (p. 7). To achieve this, it would "create policies, attitudes and values associated with a learning organization; create and maintain a learning organization through intellectual effort and by creating, accepting and practicing appropriate values; look outside the CF for knowledge; and create a well-balanced set of incentives for initiative, debate and responsible risk-taking" (p. 12). Recently, the Australian Defence Force's (2007) leadership doctrine publication (Executive Pamphlet 00.6) also introduced the idea of the "Australian Defence Force as a learning organization" as a "contemporary leadership issue", suggesting that certain military leadership behaviors "are required to adapt to changes in the environment rather than just react to them."

Peter Senge on "Learning Organization"

What does it take for traditionally hierarchical, mission-focused military forces to function as learning organizations where its people are comfortable with constant change and many "unknowns"? What are the challenges of military leadership in such "learning organizations"? The concept of a Learning Organization was popularized by Peter Senge (1990) in *The Fifth Discipline.* Senge argued that the increasing rate of change in organizational environments made it imperative for organizations to try to harness the individual and collective learning of all organizational members in order to stay competitive. He proposed a vision of organizations "where people continually expand their capacity to create the results they truly desire, where new and expansive patterns of thinking are nurtured, where collective aspiration is set free, and where people are continually learning how to learn together." (p. 3). In such organizations, all members would practice five individual and collective habits or "disciplines, which Senge labeled personal mastery, team learning, shared visioning, mental models and systems thinking (elaborated in Table 9-3).

TABLE 9-3. Senge's Five Disciplines of a Learning Organization

Discipline	Description
Personal Mastery	The practice of continually clarifying and deepening one's personal vision of a desired future, while concurrently seeing one's current reality as objectively as possible. This requires patience, self-awareness of weaknesses and growth areas, a sense of humility, and persistence to pursue self-development.
Mental models	Recognizing that all people have certain deeply-held assumptions, generalizations and biases that shape their perception of events, this refers to the practice of continually clarifying one's "unearthing", "scrutinizing" and even challenging one's mental models which is necessary for learning and change to occur.
Shared vision	This refers to a collective practice of building a shared picture of the desired future which would engender individual enrollment and genuine commitment toward a cause rather than compliant action. This provides a shared sense of purpose and helps people to make sense of the many different activities in the organization.
Team learning	This refers to the practice of listening deeply to each other (or suspending judgment or not "jumping to conclusions" or being defensive) in the process of dialogue within a group, with a genuine spirit of wanting to learn and to harness the greater knowledge and insights of the team.
Systems thinking	This is the practice of trying to see the "big picture" by establishing patterns and inter-relationships across organizational elements and events in a holistic and comprehensive way, while overcoming simplistic or linear thinking.

Source: Adapted from Senge (1990).

Senge referred to systems thinking as the "Fifth Discipline" which was most critical for integrating all four other disciplines in a learning organization. To practice systems thinking would require all to view all organizational components and events from a systems perspective with delays, feedback loops, rather than from a linear, unidirectional cause and effect perspective. It is important that people discard their old ways of thinking or their "mental models" and develop a clear sense of both their personal vision and the current reality so as to achieve "personal mastery". They should also constantly try to understand how different parts of the organization synergize (i.e., to practice "systems thinking"), and constantly build a consensus of the desired future ("shared visioning") for the organization, while also working together to learn and to achieve their shared vision (i.e., to practice "team learning"). Peter Senge referred to these as five "disciplines" as akin to "disciplines of study" where people had to constantly study and practice certain principles, ideas and methods related to these five ideas so that these could become "integrated" into their lives – like "habits".

For Senge, constant practice of the five disciplines would help overcome certain "learning disabilities" or thinking or assumptions which hinder learning and change for the organization as a whole. One such learning disability is the tendency of organizational members to see their influence only within the immediate sphere of their job or appointment and to draw clear boundaries of responsibility such that they "become their position" and fail to see how their actions (or inaction) can affect other positions in the organization. Such self-serving, self-protecting actions by each individual in the organization is as opposed to collective, creative actions needed to help the organization as a whole to change or transform itself to cope with new environmental challenges. The "I am my position" disability results in actions that focus on establishing "whose fault it is" or finding or blaming "an enemy out there" when a problem occurs, which is in turn another learning disability which Senge terms "the illusion of taking charge". This pattern of reacting to immediate problems rather than attending to long-term desired outcomes is another disability which Senge calls "fixation on events". Other organizational learning disabilities identified by Senge include "the parable of the boiled frog" which refers to the danger or adapting to gradually emerging organizational threats (akin to how frogs as cold blooded animals would adapt to gradually boiling water but would jump out if suddenly put into hot water), "the delusion of learning from experience", which occurs because much of our personal experience of outcomes are immediate rather than long-term, organization-wide ones, and the "myth of the management team" or which assumes that management teams can avoid the above learning disabilities especially when they are faced with crises or pressures.

Organizational Development, Change and Learning

Many of Peter Senge's ideas about the learning organization were built on ideas from more traditional fields like organizational development and change, and the study of organizational learning, both of which had origins in the study of group and organizational "dynamics" in the 1950s.

Organizational development is defined as "an effort (1) planned, (2) organization-wide, and (3) managed from the top, to (4) increase organization effectiveness and health through (5) planned interventions in the organization's processes, using behavioral science knowledge " (p. 9; Beckhard, 1969). In the 1970s, the terms organizational learning and organizational change were often used interchangeably with organization development. The concept of organizational learning was first introduced by the behavioral scientists, Argyris and Schon (1974, 1978, 1996), who also introduced the ideas of single versus double-loop learning, illustrated in Figure 9-1. Argyris and Schon argued that an important condition for organizational learning to occur was change in the individual members' mental models or "theory" of the organization. In their view, organizational learning required learning or cognitive change at the individual level to occur. This idea was also echoed by Senge (1990) who wrote: "organizations learn only through individuals who learn. Individual learning does not guarantee organizational learning. But without it no organizational learning occurs" (p. 139). In his vision, Senge described learning organizations as not just adapting to their environments but having the "generative" or "creative" capacities to "stay ahead" of changes in the environment. He argued that for an adaptive, learning organization, learning was more than a matter of mere survival: ". . . for a learning organization, 'adaptive learning' must be joined by 'generative learning' or learning that enhances our capacity to create" (1994, p. 14). With this view, Senge was trying to capture the ideas of single versus double-loop learning first introduced by Argyris and Schon in the 1970s. In contrast to single-loop learning which involves the detection and correction of errors within a given set of parameters or variables that typically result in incremental changes, double-loop learning involves examining the parameters or variables themselves (or the broader theory of the organization) which may result in more fundamental or long-term change or re-design.

FIGURE 9-1. Single and Double-loop learning.

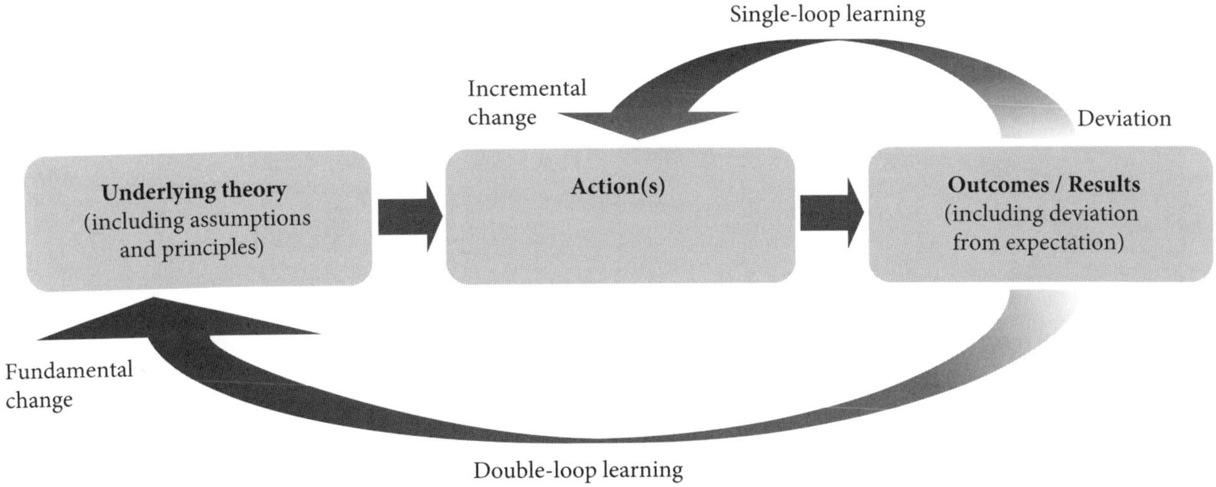

"Demystifying" the Learning Organization

Lipshitz, Friendman & Popper (2007; see also Popper & Lipshitz, 1998, 2000a, 2000b) noted that one challenge of the Learning Organization "movement" was its soft, intangible orientation, which resulted in many becoming skeptical of the idea. They suggested that one reason for this was because learning organization practitioners tended to use metaphors and heuristic tools rather than basic scientific theory, and they also tended to focus on "processes" rather than "outcomes".

Taking a different approach, Lipshitz et al., (2007) who had studied organizational learning in the Israeli Defense Forces, proposed that organizational learning can be operationalized in terms of two key aspects: Organizational Learning Mechanisms (OLMs) which are observable organizational structures and practices through which organization members interacted for the purpose of learning; and, an adaptive organizational culture that both facilitated the learning needed by organization to achieve its objectives while also taking into consideration the demands of the internal and external environment. Examples of OLMs include the conduct of after-action reviews (or AARs), and the establishment of "Communities of Practice" and coaching or mentoring networks.

A U.S. Army Captain Captain giving his tank platoon an After Action Review following a mock battle.
Source: Fritz Hoffman/Image Works/Time & Life Images/Getty Images

Lipshitz et al., (2007) added that merely introducing organizational learning mechanisms did not guarantee that organizational learning would occur or that the learning would be productive. They emphasized that such structures worked best only within a learning culture in the organization. This was because organizational learning took place in a social context that required the cooperation of others to provide undistorted information and a degree of trust and critical thinking needed for interpretation of any new information from multiple perspectives. From their research, Lipshitz et al. (2007) suggested five cultural norms that could help generate valid knowledge and lead people to act on this knowledge in organizations. These were:

- **Inquiry**: This refers to the behavior of suspending judgment and persisting in a spirit of inquiry until complete understanding is achieved. This is as opposed to being defensive or trying to "save face" at the expense of learning or improving. The key to spirit of inquiry is a positive orientation to knowledge or a "knowledge-friendly culture".
- **Issue orientation**: This means focusing one's effort to understand the specific issue at hand as opposed to being defensive when receiving negative information. It also means objectively considering the relevance of information regardless of the rank or status of the person giving or receiving this information, and not focusing one's attention on personal or interpersonal concerns.
- **Transparency**: This involves surfacing one's thoughts, actions and assumptions so that these are open to the scrutiny of others.
- **Integrity**: This involves admitting that one made a mistake when provided with compelling evidence — even if it means incurring losses in the process. Together, integrity and issue orientation help people to proceed to learn in a spirit of inquiry despite the threat of change that may come with learning.
- **Accountability**. This involves taking responsibility for one's learning and acting to implement one's lessons learned.

Case Study: U.S. Army's Transformation to a Learning Organization

What does it take for a military organization to transform itself into a learning organization — one where learning and change is the only constant? In this section, we relate the U.S. Army's organizational transformation efforts in two periods, with focus on how it attempted to influence its leadership culture. The first period began with the challenges faced by the U.S. Army after the Vietnam War to its success in Operational Desert Storm in 1990. The next period began from the end of the Cold War to the mid-1990s. The story of how the U.S. Army transformed itself after the Vietnam War is told in a video documentary entitled *All We Could Be: The Rebuilding of the Army from Vietnam to Desert Storm* (Brown, 2002). Its transformation in the second period after the fall of the Berlin Wall in 1989 is best described in the book *Hope is not a Method* by Sullivan and Harper (1996). One theme that emerges is the importance of pervasive, organization-wide cultural and intellectual changes beyond formal organizational development initiatives needed to create a learning organization.

U.S. Army's post-Vietnam "Organization Development" initiatives. In a book chapter entitled *The Death and Birth of O.D.*, Tichy and DeRose (2003) described how in the late 1970s when the many organizations failed to see the relevance of external consultant-driven organizational development to their strategies, the U.S. military played a key role in "mainstreaming" the field of organizational development by establishing an Organizational Effectiveness School at Ford Ord (see also Deaner, 1991; Roberts & Barko, 1986; Umstot, 1980). Deaner (1991) wrote that the U.S. Army's Organizational Effectiveness (OE) Program emerged in response to the turmoil in the Army during the late 1960s and early 1970s (p. 12), during the post-Vietnam War effort by the U.S. Army to transform itself into a professional military force. This School apparently infused into many of the U.S. Army's most promising lieutenant colonels and colonels the basic principles and techniques of organizational development, including the techniques of survey feedback, team building and process facilitation. Although the U.S. Army's Organizational Effectiveness program was closed down in the mid-80s for various reasons, Tichy and DeRose (2003) described how the chairman and CEO of General Electric, Jack Welch, used the U.S. Army's OE model as a framework for transforming GE by designing its leadership development facility at Crotonville and infused the basic principles and techniques of OD "into the leadership repertoire of tens of thousands of GE workers" (p. 158).

"Hope is not a method" (Sullivan & Harper, 1996). The fall of the Berlin Wall in November 1989 marked the end of the "Cold War" between the NATO allies and the Soviet Union, resulting in many military forces on both sides of the "Iron Curtain" facing pressures to downsize in the 1990s. General Gordon Sullivan, the U.S. Army Chief of Staff from 1991 to 1995 and his then-key strategic planner, Colonel Harper found themselves faced with an expected 40% reduction in staff and funding, and a military that needed to keep up with new information age technologies. In 1996, they published *Hope is not a Method* to explain how the U.S. Army dismantled its Cold War bureaucracy and transformed itself to possess generative learning capacities so that it could constantly stay ahead of new challenges while maintaining a focus on current operations. Part of the change required a "new theory" or "military doctrine" for the Army's business. General Sullivan described how he gathered the Army's senior leadership to develop a common framing of the Army challenges and a consensus on the need for change. The Army's senior leadership then developed a new unified concept which it "tested" in various experimental settings. To effect the organization change, General Sullivan used his credibility and relationship with soldiers which he demonstrated through personal involvement, speeches, video-recordings, and personal visits. He also managed the expectation and belief that the downsizing would occur in as fair and equitable a manner as possible, and identified the shared Army values and concerns with the entire Army as a "family" and used both reason and emotional appeals to connect with soldiers. Sullivan & Harper (1996) also emphasized the importance of teamwork and "distributing leadership" in the organization: "Distributing leadership empowers people to act within boundaries of acceptable behavior as defined by the values, vision, and strategy of the organization. Distributing leadership, however, also means distributing responsibility. . . . We want all our people to be leaders in the sense that we want each to feel an acute sense of responsibility for the performance of the whole organization" (p. 119). Drawing on their experience of transforming the U.S. Army, they identified the various tasks for leaders in transforming organizations — many of which were related to ideas articulated by Senge on the Learning Organization:

- Leaders must focus on the people aspect or human dimension of change, and not just attend to new ideas or technologies.
- Leaders need to both "scan and focus". They must maintain a broad personal outlook of the organization and its environment while at the same time communicate what they are seeing and learning to others in a purposeful way.
- Leaders need to design long-term, "strategic architecture" for their organizations which are anchored on "cornerstones" of values and a vision.
- Leaders need to be able to "see patterns where others cannot". They must also have the courage to make decisions — often with incomplete information — and to act quickly when required (rather than to "hope" for a solution).
- Leaders must lead for a learning organization, where the important behaviors include networking, influencing, and inspiring others rather than only directing others.
- Leaders must have the ability to teach and develop their subordinates, and not just to focus on task performance or results.

MILITARY FORCES ON LEADERSHIP AND ORGANIZATIONAL CHANGE

U.S. Army on "Improving the Organization

The role of military leaders in organization development or "improving the organization" was formally introduced into the U.S. Army's lexicon in 1999. The1990 edition of the U.S. Army Field Manual (FM) 22-100 defined leadership as: "the process of influencing others to accomplish the mission by providing purpose, direction, and motivation". Its 1999 edition defined leadership as "the process of influencing people by providing purpose, direction, and motivation while operating to accomplish the mission and *improving the organization*" (italics added). The Preface to the 1999 FM22-100 manual mentioned that this edition would have a focus on "improving people and organizations for the long term", and, that "improving" would be introduced as a category of "leader actions" that would apply to all levels of army leadership. "Improving actions" was explained in 1999 FM22-100 with an example as follows: "The motor sergeant's job is not complete when the last vehicle clears the gate. While getting the job done is key, the Army also expects him to do far more than just accomplish the day's work. Army leaders also strive to improve everything entrusted to them: their people, facilities, equipment, training, and resources. There will be a new mission, of course, but part of finishing the old one is improving the organization" (p. 1–5). The manual went on to describe how it would expect the motor sergeant to also conduct an after action review (AAR) with his work team to discover gaps in their individual or collective actions, and then to conduct training or developmental counseling to improve their future performance. The manual explained that: "By doing these things, the motor sergeant is creating a better organization, one that will work smarter the next time. His example sends an important message. The soldiers see their leader look at his own and the organization's performance, evaluate it, identify strong areas to sustain as well as mistakes and shortcomings, and commit to a better way of doing things. These actions are more powerful than any lecture on leadership" (p. 1–6).

Other forces

Today, most contemporary military leadership doctrines include organizational development or "improvement" as one aspect required of military leadership at all levels. In its 2005 leadership doctrine, the Canadian Forces defined effective leadership as "Directing, motivating and enabling others to accomplish the mission professionally and ethically, while developing or improving capabilities that contribute to mission success" (p. 5). The Singapore Armed Forces doctrine defines leadership as "a process of influencing people to accomplish the mission, inspiring their commitment, and improving the organization" (see Chan, Singh, Ramaya & Lim, 2005). The Australian Defence Force's 2007 doctrine publication 00.6 introduced the idea of "ADF — A Learning Organization" as a "contemporary leadership issue" and suggested certain behaviors expected of its leaders including: "Encouraging critical and creative thought; taking appropriate risks to explore opportunities; creating a climate conducive to individual and team learning; encouraging learning from mistakes rather than a culture of blame; providing incentives for learning, experimentation and innovation" (p. 5.2).

SUMMARY

This chapter has attempted to discuss the challenges of leadership in modern military organizations, which, are on one hand "traditionally hierarchical", and, which on the other hand increasingly have to operate as "learning organizations". The chapter began by introducing key organizational concepts such as organizational structure and culture, and how leadership affects and is in turn influenced by these. From a traditional, hierarchical perspective, military leaders need to understand the different tasks and challenges of leading at different levels of an organizational hierarchy and also the challenges of leading in a formal organizational context. The U.S. and Canadian forces identify different number of leadership "levels" in their doctrines; but both are based on the research-based Stratified Systems Theory developed by Jacobs and Jacques (1987, 1990, 1991). The chapter then notes how organizational structures are changing in the sense of becoming less rigid and inflexible or "boundaryless", and that such weakening of organizational structures may demand more rather than less leadership in organizations. Some have even called for a more "distributed" concept of leadership as opposed to one that is centered on individual leaders.

Today, it is common to find military forces adopting the ideas of a "learning organization", or emphasizing the importance of organizational learning. The concept of a learning organization popularized by Senge (1990) in *The Fifth Discipline*, which built on foundational knowledge from the fields of organization development and organizational learning. Senge's Learning Organization calls for more systems thinking along with other practices like personal mastery, shared vision, team learning and mental models, which together, are critical for overcoming various "learning disabilities" in organizations. Israeli psychologists, Lipshitz et al. (2007), propose that the soft, intangible concept of a Learning Organization can be operationalized in terms of two key aspects: "Organizational learning mechanisms" (OLMs) and an organizational culture that facilitates learning.

A brief description of the U.S. Army's efforts to transform itself from the post-Vietnam War era to the 1900s attempted to highlight the importance of pervasive, organization-wide cultural and intellectual changes — beyond formal organizational development initiatives — that are needed to create organizations with a capacity to grow and cope with highly complex, dynamic and uncertain environments. Today, some military leadership doctrines also include ideas from organizational development or "improving the organization" as a set of competencies required at all levels of military leadership. Readers who are interested in the nature of leadership required learning military organizations may refer to *Hope is not a Method* by Sullivan and Harper (1996), which shares many tips for leadership in constantly transforming military organizations.

Key Concepts

- Organizational structure
- Organizational culture
- Organizational climate
- Levels of Leadership
- Timespan of responsibilities
- Direct leadership
- Indirect leadership
- Organizational leadership
- Stratefied Systems Theory
- Strategic leadership
- Learning Organization
- Peter Senge's "Five disciplines"
- Organizational Learning
- Organization Development
- Single vs. double-loop learning
- Boundaryless organization
- Organizational learning mechanisms
- Distributed Leadership

Questions

- Is it useful for junior military leaders to have some understanding of why their senior leaders may have to "lead" in different ways from them? Or, should leaders at each level only understand leadership at that level? Why?
- Do you believe that military forces increasingly need to operate like "learning organizations" as described by Senge? Or, do you believe that the idea of a learning organization or organizational learning does not apply to military forces? Why?
- What are the tensions and challenges of operating and leading in military organizations that are both "traditionally hierarchical" and also "learning organizations"? How should military forces' training and education systems prepare leaders to lead at different levels of the organization, and in organizations that are both "traditionally hierarchical" and yet "learning organizations"?

References

Argyris, C., & Schön, D.A. (1974). *Theory in practice: Increasing professional effectiveness*. San Francisco: Jossey-Bass.

Argyris, C., & Schön, D.A. (1978). *Organizational learning: A theory of action perspective*. Reading, MA: Addison-Wesley.

Argyris, C., & Schön, D.A. (1996). *Organizational learning II: Theory, methods and practice*. Reading, MA: Addison-Wesley.

Ashkenas, R., Ulrich, D., Jick, T., & Kerr, S. (1995). *The Boundaryless organization: Breaking the chains of organizational structure*. San Francisco, CA: Jossey-Bass.

Australian Defence Headquarters (2007). *Leadership in the Australian Defence Force (ADDP 00.6)*. Director Centre for Defence Leadership Studies, Defence Publishing Service, Canberra, ACT.

Beckhard, R. (1969). *Organization development: Strategies and models*. Reading, MA: Addison - Wesley.

Brown, F.J.. (2002). *All We Could Be: The Rebuilding of the Army from Vietnam to Desert Storm*. Thousand Oaks, CA: First Person Productions.

Burke, W.W., & Litwin, G. (1989). A causal model of organizational performance. In J.W. Pfeiffer (Ed.), *The 1989 annual: Developing human resources*. San Diego: University Associates.

Canadian National Defense (2001). *Canadian officership in the 21st century (Officership 2020): Strategic guidance for the Canadian Forces officer corps and the officer professional development system. Canada: Department of National Defense*. Retrieved from http://www.cda-acd.forces.gc.ca/pub/2020/officer/doc/off-2020-s-eng.pdf on 10 January 2010.

Canadian National Defense (2005a). *Leadership in the Canadian Forces: Doctrine*. Kingston, ON: Canadian Forces Leadership Institute, Canadian Defense Academy.

Canadian National Defense (2005b). *Leadership in the Canadian Forces: Conceptual Foundations*. Kingston, ON: Canadian Forces Leadership Institute, Canadian Defense Academy.

Canadian National Defense (2007a). *Leadership in the Canadian Forces: Leading People*. Kingston, ON: Canadian Forces Leadership Institute, Canadian Defense Academy.

Canadian National Defense (2007b). *Leadership in the Canadian Forces: Leading the Institution*. Kingston, ON: Canadian Forces Leadership Institute, Canadian Defense Academy.

Chan, K.Y., Singh, S., Ramaya, R., & Lim, K.H. (2005). *Spirit and system: Leadership development for a third generation SAF*. SAFTI Military Institute, Singapore: Pointer Monograph.

Collins, J. (2001). *Good to great: Why some companies make the leap . . . and others don't*. NY: HarperCollins Publishers

Collins, J., & Porras, J. (1996). *Built to last*. New York, NY: Harper Collins,

Deaner, C.M.D. (1991). The U.S. Army Organizational Effectiveness Program: A Eulogy and Critique. *Public Administration Quarterly, 15(1)*, 12–31.

Fayol H. (1916). General and Industrial Management. *Bulletin de la Société de l'Industrie Minérale, 10*, 5–164.

Hunt, J.G., Dodge, G.E., & Wong, L. (Eds.), *Out-of-the-box-leadership: Transforming the twenty-first-century soldier*. Stamford, CT: JAI Press.

Jacobs, T.O., & Jaques, E. (1987). Leadership in complex systems. In J. Zeider (Ed.), *Human productivity enhancement*. (pp. 7–65). New York, NY: Praeger.

Jacobs, T.O., & Jaques, E. (1990). Military Executive Leadership. In K.E. Clark et al. (Eds.), *Measurement of leadership*. West Orange, NJ: Leadership Library of America, Inc.

Jacobs, T.O., & Jaques, E. (1991). Executive leadership. In R. Gal & A.D. Manglesdorff (Eds.), *Handbook of military psychology*. Chichester, England: Wiley.

Jaques, E. (1989). *Requisite organization*. Arlington, VA: Cason-Hall and Co. publishing.

Jaques, E., Gibson, R.O. & Issac, D.J. (1978). *Levels of Abstraction in Logic and Human Action: A Theory of Discontinuity in the Structure of Mathematical Logic, Psychological Behavior and Social Organization*. London: Heinemann.

Jacques, E., & Stamp, G. (1995). *Level and type of capability in relation to executive organization (Research Note 95-13)*. Alexandria, VA: Army Research Institute for the Social and Behavioural Sciences.

Lipshitz, R., Friedman, V.J., & Popper, M. (2007). *Demystifying organizational learning*. Thousand Oaks, CA: Sage.

Morgan, G. (1986). *Images of organization*. Thousand Oaks, CA: Sage

Popper, M., & Lipshitz, R. (1998). Organizational Learning Mechanisms: A Cultural and Structural Approach to Organizational Learning. *Journal of Applied Behavioral Science, 34*, 161–78.

Popper, M., & Lipshitz, R. (2000). Organizational learning: mechanisms and feasibility, *Management Learning, 31(2)*, 181–96.

Popper, M., & Lipshitz, R. (2000). Installing mechanisms and instilling values: the role of leaders in organizational learning. *The Learning Organization, 7(3)*, 135–144.

Robbins, S.P. (1987). *Organization Theory: Structure, Design and Application (2nd ed.)*. Englewood Cliffs, NJ: Prentice-Hall,

Roberts, B.J., & Barko, W.F. (1986). Organizational development in the U.S. Army: A conceptual case analysis. *Public Administration Quarterly, 10(3)*, 325–335.

Schein, E.H. (1985). *Organizational culture and leadership*. San Francisco, CA: Jossey-Bass.

Schneider, B. (2000). The psychological life of organizations. In N.M. Ashkanasy, C. Wilderom, & M.F. Peterson (Eds.), *Handbook of organizational culture and climate*. Thousand Oaks, CA: Sage.

Senge, P. (1990). *The fifth discipline: The art and practice of the learning organization*. New York: Doubleday.

Shamir, B. (1999), Leadership in boundaryless organizations: disposable or indispensable? *European Journal of Work and Organizational Psychology*, 8, 49–71.

Shamir, B., & Ben-Ari, E. (1999). Leadership in an open army? Civilian connections, interorganizational frameworks, and changes in military leadership. In J.G. Hunt, G.E. Dodge, & L Wong (Eds.), *Out-of-the-box-leadership: Transforming the twenty-first-century soldier* (pp. 15–40). Stamford, CT: JAI Press.

Sullivan, G.R., & Harper, M.V. (1996). *Hope is not a method: What business leaders can learn from America's Army*. New York: Broadway Books.

Taylor, F.W. (1911). *The Principles of scientific management*. New York: Harper Bros.

Tichy, N.M., & DeRose, C. (2002). The death and rebirth of organizational development. In S. Chowdhury et al. (Eds.), *Organization 21C: Someday all organizations will lead this way*. Upper Saddle River, NJ: FT Press.

U.S. Army (1983). *Military leadership* (Field Manual 22-100]. Washington, D.C.: Headquarters, Department of the Army.

U.S. Army (2006). *Army Leadership* (Field Manual 6-22]. Washington, D.C.: Headquarters, Department of the Army.

U.S. Army (1987). *Leadership and Command at Senior Levels* (Field Manual 22-103]. Washington, D.C.: Headquarters, Department of the Army.

U.S. Army (1987). *Executive Leadership Levels* (Pamphlet 600-80]. Washington, D.C.: Headquarters, Department of the Army.

Umstot, D.D. (1980). Organization Development technology & the military: A Surprising merger? *Academy of Management Review, 5(2)*, 189–201.

Wong, L., Bliese, P., & McGurk, D. (2003). Military leadership: A context-specific review. *The Leadership Quarterly, 14(4)*, 657–692.

Zaccaro, S.J. (1996). *Models and theories of executive leadership: A conceptual/empirical review and integration*. Alexandria, VA: U.S. Army Research Institute for the Behavioral and Social Sciences.

Zaccaro, S.J., & Klimoski, R.J. (2001). The nature of organizational leadership: An introduction. In: S. J. Zaccaro & R. J. Klimoski (Eds.), The nature of organizational leadership (pp. 3–41). San Francisco, CA: Jossey Bass.

Yerkes, R.M. (Ed.) (1921). Psychological examining in the United States Army. *Memoirs of the National Academy of Sciences, 15*, 1–890.

Zaccaro, S.J. (2001). *Nature of Executive Leadership: A Conceptual and Empirical Analysis of Success*. Washington, D.C.: American Psychological Association.

10

Leading in the 21st Century Profession of Arms

Source: CyberPioneer, Ministry of Defence, Singapore

OVERVIEW

The previous chapter discussed the unique challenges and nature of 21st century military leadership in the context of its traditionally hierarchical yet contemporarily learning organizational forms. This chapter focuses on the challenges of leadership in context of the military both as *a professional career and institution*. It begins with a general discussion on the concept of "professions" which is followed by a discussion on the idea of "the military as a profession". The origins and features of both concepts are introduced. Next, the chapter provides a framework for considering how the profession of arms is changing in the 21st century — both organizationally as a social institution, and individually in terms of a military career. This chapter takes a more multidisciplinary approach by drawing ideas from the psychology, management, sociology and the humanities. Concepts such as "unlimited liability", "post-modern military" and "civil-military relations" are introduced to clarify the unique nature of the military profession today.

Just as Chapter 9 described how military organizations are changing in form to become more "boundaryless", this chapter suggests that military professionals are also challenged by the boundaryless nature of careers in a 21st century work context where people entering the workforce increasingly desire more self-direction and autonomy in their careers and are less prepared to establish long-term relationships with their employing organizations. In view of more boundaryless military organizations and career forms, military leaders at all levels will increasingly have to play an integrative "institutional role" by actively strengthening military culture through active involvement in socialization and professional military training and education initiatives, and participating in efforts to clarify and reinforce military values and ethos. They also need to clarify and strengthen civil-military relations, and influence intra-organizational, professional effectiveness. It is also suggested that military leaders must also complement the typically transactional, bureaucratic nature of human resource practices with a strong element of authentic, transformational and values-based leadership. These ideas are also discussed in context of various broader strategies being adopted by military forces and nations to deal with the challenges facing the profession of arms in the 21stcentury.

WHAT IS A "PROFESSION"?

Before discussing the modern concept of "professions", it is useful to clarify the difference between the concepts of "a profession", "a professional" and "professionalism". This chapter will focus on the current, conventional understanding of the term as used in discussions on work and careers. In this chapter, a "profession" refers to a distinct kind of work, vocation or career that demands a highly specialized and complex body of formal knowledge and expertise which is typically acquired through formal training and/or education, and that must meet certain standards of qualification and/or professional certification. In contrast, the word "professionalism" refers to an attitude or orientation that any person — not only those in formal professions — may adopt their work.

The term "a profession" refers to the formal or informal group or category of professionals. Different professions may differ in the degree of formalization in different states. The most formal professional organizations would have some degree of state-level laws to regulate the profession. These would in turn have some formal system of certification or qualification, continual professional education programs, and codes of conduct or guidelines to regulate the individual members' professional conduct. The concept of "professions" is one that constantly evolves with the meaning of work and ways in which work is organized in different societies. In most societies today, the most "prototypical" professions would include doctors, lawyers, engineers, accountants and teachers. It is also useful to distinguish between the "profession" as an organization or grouping of people, versus a "professional" as an individual member of a profession. A "professional" therefore refers to a person who is engaged in one of these types of professions.

A Brief History of "Professions"

The modern, well-accepted concept of "professions" emerged during the Middle Ages in 12th century Europe, when the church and state authorities started to regulate the training and actions of various categories of workers. During this period, "higher" professions referred to certain types of work like lawyers, physicians, university professors while "lower" professions included surgeons and notaries (i.e., persons appointed by authorities to witness certain transactions). By the 19th century, the idea of professions had evolved to become closely linked to social status with terms like "gentlemanly professions" and "learned professions" being used in England, and "intellectuals" being used in Western Europe. This period also witnessed many European governments implementing laws to regulate certain professions. By the early 20th century, the relationship between professional and social status declined so that professionalism began to depend more on the mastery of a certain expertise and body of knowledge, the establishment of clear codes of conduct and the extent of self-regulation within the profession.

Today, most scholars agree that a profession is a distinct kind of work. In this context, Friedson (2001) defined "professionalism" as an official recognition that the occupation used in its work a complex body of formal knowledge and skill, including abstract concepts or theories that required the exercise of a considerable amount of discretion. Today, "professions" are those vocations recognized as possessing a high degree of specialized knowledge or expertise. Professions are expected to specify clear standards for joining such vocations, and, for the quality of education provided to its students. Fundamental to the idea of professions is the question of how certain highly specialized knowledge or expertise is generated, controlled and regenerated or transferred to future generations.

Ways to Think about "Professions"

Professions as a distinct type of work. Some of the earliest academic efforts to study professions began in the last century where the main approach was to identify certain traits or common features that could characterize all professions and differentiating them from non-professions. For example, in a paper attempting to evaluate whether social work could be treated as a "profession", Flexner (1915) identified certain features that he thought should distinguish professions from other occupations. Another noted attempt at identifying 'traits" of professions was by Carr-Sanders and Wilson (1933). This "trait approach" to professions was not successful because of the changing nature of professions, and because different professions were organized differently across different countries. For example, in the early 1900s, most professions (excluding the military) were organized in terms of independent practices with some type of collective professional association. It was only in the second half of the 20th century that more and more professionals started to work in corporate settings as salaried workers rather than as independent practitioners.

In a literature review, Millerson (1964) concluded that there was very little consensus on the common features of professions, but suggested six essential characteristics of professions as follows: (1) professions possess and apply specialized skills that are based on theoretical knowledge; (2) specialized education and training is required to master professional skills; (3) the competence of professionals is maintained and ensured via tests and examinations; (4) professions should possess a code of conduct to maintain a high standard of professional integrity; (5) professions perform a service to meet the needs of the public; (6) professional members are organized by some formal association. In another review, Eraut (1994) emphasized the point that professions evolved with changes in societies and culture and identified three "central features" of the "ideology of professionalism" as follows: (1) a specialist knowledge base, (2) autonomy, and (3) service.

Writing on the *military* profession, Huntington (1957) identified three defining characteristics of a profession which he used as a basis to discuss the question of "who is a military professional" (see subsequent section discussing this issue). These were:

- **Expertise:** This refers to the highly specialized knowledge that is only acquired through prolonged education and experience, and that requires continual study and practice of the profession.
- **Responsibility:** The attitude that one practices his or her profession in the service of society to hold certain moral principles in dealing with laypersons.
- **Corporateness:** This refers to the sense of unity among the professions and separateness from those who are not members of their profession, which may arise partly because of the lengthy training necessary to achieve professional competence, and is also supported by the establishment of professional associations and societies.

Huntington's three characteristics of a profession are expertise, responsibility and corporateness.

Source: CyberPioneer, Ministry of Defence, Singapore

Professionalization as an on-going "process". Rather than emphasizing the uniqueness of professions relative to other kinds of work, some scholars think of professionalization as a dynamic process where various groups in society strive toward achieving certain characteristics (e.g., standards, exclusivity, control, monopoly, autonomy of self-regulation, etc.) or outcomes (e.g., societal trust). This view is supported by the origins of modern professions which emerged in an age of competitive capitalism in the late 18th and early 19th centuries. Adopting this view, the German sociologist, Max Weber, for example, described how professions were the result of attempts to monopolize or to exclude other occupations in meeting certain societal needs through education and credentials. In the 1970s a more skeptical view emerged that argued that professions are "occupational groups that can successfully convince the public of the societal necessity of their skills and the 'rightness' of their authority" (Sanders and Lyon, 1976). In less political terms, some sociologists like Houle (1981), Vollmer & Mills (1966) describe professions as a "dynamic process" of professionalization in which members must constantly reflect on how well the profession was achieving its goals in various domains: Conceptual or theoretical, Performance, and Collective identity aspects. The organizational learning theorist, Donald Schon (1983, 1987), also stressed how professional expertise and knowledge was increasingly based not only on scientifically established knowledge applied to practical problems but also on the tacit knowledge held by individuals and organizations.

A developmental perspective: How professions come about. MacDonald & Ritzer (1988) proposed a lifespan view of professions, that over time, professions strived to carve out the area they wished to control as much as possible of the work closely related to this area without associating too closely with subordinate groups. Members of occupations hoping to attain or retain professional status developed certain "schools of thought" that supported their claim of exclusive jurisdiction over particular tasks. Typically, professions began with full time service in their area of professional responsibility. The next step was the establishment of a training school and the formation of a professional association. With these structures, the profession then mades political efforts to get legal recognition and to protect their "turf". In doing so, they also established formal codes of ethics to ensure standards of practice and rules to protect their "clients".

How professionals think: The psychology of "professionalism". The above approaches were essentially sociological — they treated professions as a category or sector in any society. At the individual level, social and organizational psychologists had also attempted to study professionalism in terms of a certain set of "attitudes" or "beliefs". Richard Hall (1968; 1969), for example, developed an attitude scale to measure the degree of professionalism among practitioners of various occupations in terms of five dimensions. He believed that the essential beliefs of professionals included: (1) the use of the professional organization as a major point of reference in one's professional life, (2) a strong belief in "public service" or a sense of "civic duty", (3) a belief in self-regulation or standards and conduct within the profession, (4) having a sense of "calling" to the field, and, (5) a feeling of professional "autonomy" in how one operates or makes decisions as a professional. Similarly, psychologists Kerr, Von Glinow,

& Schriesheim (1977) argued that professionalism should be considered a multidimensional construct, and identified five specific attitudinal dimensions that appeared to characterize "professionalism". They labeled these attitudes or beliefs: (1) a desire for professional autonomy or the independent right to make professional judgments and decisions, (2) a sense of commitment to one's work and to the profession as a whole, (3) identification with the profession and with other professions as a unique category of work, (4) professional ethics or a sense of obligation to serve others without concern for self-interest or becoming emotionally involved with the client, and (5) belief in collegial or peer-level maintenance and regulation of standards and conduct. Note: Both lists do not include "expertise" because they treat it as an ability ("can do") rather than as an attitudinal ("will do") construct.

Prototypical features of professions. The above discussion has emphasized that the concept of a "profession" or "professionalism" can be understood at two levels — namely the individual and organizational levels. This is important in the subsequent discussion on the challenges facing the military profession in the 21st century which presents these "challenges" at both the societal-organizational and individual-career levels. It may also be more useful to think of "prototypical features" as opposed to fixed "traits" of professions or professionals. Adapting Huntington's (1957) framework of three defining traits of professions, the prototypical features of professions can be organized into three clusters as follows (see Figure 10-1 for depiction of these features):

- **Professional expertise** or the specialized knowledge and skills of the profession or the professional. This includes the extent of specialization, codification of expertise and the basis of the profession or professional's expertise in theory and/or research. This "cluster" of features also includes the extent of the professional's perceived autonomy in employing the expertise, and the extent in which the expert knowledge is generated, its quality controlled, and how knowledge transferred to future professionals.

FIGURE 10-1. Prototypical features of "professions".

Expertise

- **Specialization**, including uniqueness, power and "trust" required in its use.
- **Quality**, including basis in theory/research, control of standards and relevance.
- **Autonomy in practice**, including extent of independent judgment.
- **Knowledge**, including its generation, codification, retention, dissemination and education.

Organization

- **Formalization of structures** (e.g., having an Association or society, rules, codes of conduct, and the means of enforcing these, etc.)
- **Strength of Culture**, e.g., shared values, practices, etc.
- Informal peer or collegial social **networks and broader community** (e.g., families, alumni, etc.)

Intrinsic Motivation & ethos

- Sense of **duty and commitment to service**.
- Sense of **calling and obligation** (to the profession and its purpose and ideas).
- Sense of **accountability**, **ethics**, **responsibility** in professional practice.

- **Professional organization**, which in its most ideal form is represented by formal associations or societies with certain rules, regulations, codes and values that members uphold as part of their professional conduct. This can also include informal networks with activities that professional members participate in, identify with or feel committed. It also includes the extent of cohesion and "esprit de corps" among members in the profession.
- **Professional motivation and ethos**, which should include an intrinsic sense of duty, service to society or the public, accountability, responsibility and obligations, and a sense of "calling" to the field — both of the collective and at the individual levels.

THE MILITARY AS A "PROFESSION"

One of the best accounts of the origins of the modern concept of the "military profession" can be found in Huntington's (1957) landmark study, *The Soldier and the State*. In today's context, the idea of sovereign nation-states and the role of military forces being established to protect these states can be traced to the signing of the Treaty of Westphalia in 1648. Until the early 1800s, military officers in European countries were drawn mainly from the ruling classes, the nobility or aristocrats, where it was common for both military commissions and promotions to be obtained by purchase. Hackett (1963) suggested that this practice existed then because of the belief that only those born into higher-class families could possess the skill and judgment that was required for success in battle. Huntington (1957) stated that the idea of military science as a distinct domain of knowledge was virtually non-existent prior to 1800. In his view, the military officers before the 1800s were best described as either mercenaries motivated by profits, or amateur aristocrats who pursued military officership for "honor and adventure". It was during the Napoleonic Wars from 1799 to 1815 that military officers in Europe began to develop techniques, values, and standards for the military to become a "profession". In 1806, after suffering a humiliating disaster at the dual battles of Jena-Auerstadt, the Prussians re-organized both their General Staff and their system of military education. In 1808, Prussia opened its officer corps beyond aristocrats, allowing all citizens to join on the basis of education and merit. In 1810, the Prussian general Gerhard von Scharnhorst established the "Kriegsakademie" in Berlin, which was designed to be a military university dedicated to studying the science of war. The success of the Prussian army over French in 1870 convinced many other European countries of the value of a more professional officer corps and by 1900, most military forces had officers who were professionals by virtue of their education, training, and competence.

Why Did Military Forces "Professionalize"?

Huntington's thesis. From the above account, one could argue that historically, the idea of the military profession emerged for primarily for reasons concerned with expertise and efficiency.

However, given the political role of military forces in protecting national sovereignty, it is difficult to disentangle such arguments from other more sociopolitical interpretations of why military forces became more "professionalized" in various states. In *The Soldier and the State* which he had subtitled "The theory and politics of civil-military relations", Huntington (1957) provided a political argument — that military professionalism was the key to civilian control over the military. Huntington argued that with professionalism, military forces themselves would focus on enhancing military efficiency while recognizing subservience to the state; that this was better than for the government to impose civilian values and directives on them, which might impair their efficiency. This emphasis on public service was captured in the farewell speech that General Douglas MacArthur (1962) gave to West Point cadets: "Yours is the profession of arms, the will to win, the sure knowledge that in war there is no substitute for victory; that if you lose the nation will be destroyed, that the very obsession of your public service must be Duty, Honor, Country". Huntington's thesis was also reflective of the field of study called "civil-military relations" that emerged at about the period of the Second World War.

"**Civil-military relations**". "Civil-military relations" is a multidisciplinary field of study that is concerned with the relationship between the military services of a state with both its civilian (including political) authorities and with the larger civil society. The idea of civilian control of the military is a basic characteristic of democratic states. This means that part of being a democracy today is that there exists a clear distinction between political and military responsibilities, and that the military is clearly subordinate to its civilian, political leadership. In modern democracies, the military should not interfere in politics and should not influence the making of national policy. While political scientists focus on the political-military aspect of civil-military relations, sociologists would tend to focus on the relationship or gap between civil society and the military. The sociological concerns include, for example, the gap between military and civil society attitudes toward national security issues and public confidence and trust in the military. Such concerns differ depending on the system of military organization – for example, countries that employ universal conscription where it is compulsory for all sons to enlist for service have to manage perceptions of public confidence in the safety of military training, in the fairness in which military leaders are selected or promoted, and the extent to which military service affects the personal lives and livelihoods of families, etc. Broader sociological aspects of civil-military relations include, for example, the relationship between the military and other government and non-governmental organizations, and the media which shapes the public image and support for military service.

Who is a Military Professional?

Huntington (1957) also argued for an exclusive view on the question of "who is a military professional". He recognized that while modern military forces contained many types of specialists such as uniformed doctors, dentists, lawyers, pharmacists, etc., it was only those who "manage violence" who were true military professionals. The first sentence of *The Soldier and the State*, states, "The modern officer corps is a professional body and the modern military officer a professional man" (p. 1). Huntington also added, "His commission is to the officer what his license is to a doctor" (p. 16). He argued that although certain specialists in the military (e.g., doctors, lawyers, etc.) might have expertise necessary to help achieve military objectives, they were not competent to "manage violence". In his view, they belonged to the officer corps in an administrative capacity, but were not part of the professional body of the officer corps. Huntington's (1957) thesis would also exclude non-commissioned officers as members of the profession of arms.

Today, most modern military forces would reject an exclusive view of the military profession. They instead attempt to articulate certain shared or prototypical features that they expect members of the profession to display or live up to. For example, in describing the *Profession of Arms in the United States*, the U.S. Army (2005) states, "A final aspect that distinguishes the American profession of arms is the professionalism of its officers and noncommissioned officers. Both are given considerable authority early in their careers. Both are expected to exercise initiative to identify and resolve unforeseen circumstances. Both are developed through a series of schools that equips them for greater responsibilities as they are promoted" (p. 1–11).

At the turn of the new millennium, the Canadian Forces debated on the issue of "who is a member of the military profession" in an effort to clarify its professional ethos after an incident where its peacekeepers in Somalia were found to have abused and killed local civilians. Today, the Canadian Forces (2003) takes an inclusive approach to the profession of arms by articulating shared beliefs including "the concept of service before self", "the lawful, ordered application of military force" and the "acceptance of the concept of unlimited liability", and also the possession of "a systematic and specialized body of military knowledge and skills acquired through education, training and experience" that members of the profession are expected to "competently and objectively in the accomplishment of their missions". It states in its profession of arms manual entitled *Duty with Honor*: "All regular force members of the CF, regardless of rank, are members of the profession of arms. Although not necessarily on full-time service, primary reserve members are an essential component of the nation's military capability and meet the criteria, and thus are accorded professional status. On active duty, they assume the status and identity of full-time military professionals" (p. 9).

Unique Features of the Military Profession

"Unlimited Liability" of members of the Profession of Arms. In *The Profession of Arms*, the British military historian, General Sir John Hackett, provided a concept to understand the unique nature of the military as a profession — one that military professionals could better identify with at the individual level. Hackett (1963) argued that the concept of "unlimited liability" in defence of national interests was what distinguished members of the military profession from other professions: "The essential basis of military life is the ordered application of force under an unlimited liability. It is the unlimited liability which sets the man who embraces this life somewhat apart. He will be (or should be) always a citizen. So long as he serves he will never be a civilian" (p. 222). The Canadian Government's (1997) *Report of the Somalia Commission of Inquiry*, attempted to explain further the uniqueness of the military profession: "the military allows for the lawful killing of others in the performance of duty. Moreover, the responsibility of military leadership permits the sacrifice of soldiers' lives in order to achieve military objectives. The stark and brutal reality of these differences from normal society has traditionally been a distinguishing feature of military life, contributing to a sense of separateness — even superiority — in relation to the civilian population". Writing for a British Broadcasting Corporation website, the military historian Professor Richard Holmes noted, "Someone who becomes a soldier is crossing a legally defined boundary. A soldier gives up some individual rights (such as the right to withdraw his labor), accepts collective standards which contribute to the common good, and undertakes, in the last analysis, to kill or be killed for a purpose in which he may have no personal interest".

Today, the concept of "unlimited liability" appears in various military doctrines. For example, the first chapter of the U.S. Army's (2005) Field Manual 1 entitled *The Army and the Profession of Arms*, states: "Individual members of the profession of arms are distinguished from those of other professions by the "unlimited liability" they assume in their oaths of office. While members of some professions engage in dangerous tasks daily, only members of the Armed Forces can be ordered to place their lives in peril anywhere at any time. The obligations they undertake, risking life and well-being for the greater good, are in many ways extraordinary" (p. 1–11). Likewise, the British Army attempted to codify the "sacred" nature of unlimited liability in its doctrine on the moral component of military effectiveness entitled *Soldiering: The Military Covenant*. The doctrine states: "All British soldiers share the legal right and duty to fight and if necessary, kill, according to their orders and an unlimited liability to give their lives in so doing. This is the unique nature of soldiering"" (p. 1–1; see also Milham, 2010).

Military as both a bureaucratic organization and a professional institution. Professions in society vary in the degree of formalization of their system of organization and governance. To the extent that each state has only one military organization, one of the most unique features of the military profession is the fact that the military is concurrently a bureaucracy organized for efficiency and a professional institution responsible for the professional effectiveness (including expertise and ethics) of its members. Snider (2002, 2005) highlighted how the dual natures of the military — as a bureaucratic organization and as profession — provided an inherent tension. In his view, the bureaucratic nature of the military organization tended to predominate over its professional aspect. Snider (2005) described how the U.S. Army had stopped studying itself as a profession for a 30-year period from the 1970s to 2000: Until the 1970s, Huntington's (1957) thesis (that the officer corps is the professional body and the military a professional man) implicitly influenced officer and leadership development. However, the emergence of "organizational theories" in the late 70s and 80s started to permeate U.S. leadership doctrine to the extent that by 1999, the U.S. Army's leadership manual (Field Manual 22-100) "failed completely to recognize the Army as a profession, using throughout the manual the generic noun 'organization' to describe the Army" (p. 7).

This "tension" between professions and bureaucracies has also been noted in the study of careers and work at the societal level. The organizational strategist, Rosabeth Moss Kanter (1989) described how at the societal level, careers could be classified into three different forms - each with its own logics and implicit motivations. She argued that while professional careers were primarily defined by expertise and reputation, bureaucratic jobs or careers were defined by a "logic" of advancement and security. The third form of career in society is the entrepreneurial one, defined by a logic of value-creation. For the purpose of this chapter, it is especially useful to note that bureaucratic mindsets can act as a threat toward military professionalism — especially when individuals see the military as nothing more than a job as opposed to a profession. Dominance of a bureaucratic mindset over a professional one could also result in careerism (vs. professionalism) and managerialism (vs. leadership) in the military (Gabriel & Savage, 1976; Moskos et al., 1986).

Military Culture. In an article entitled *An uninformed debate on military culture*, Snider (1999) argued that what made military culture unique was that its central elements "derive from the purpose or tasks for which societies raise, support, and maintain modern militaries, for instance, waging war on behalf of the nation-state and, if needed, enforcing domestic order" (p. 119). Snider identified four key elements of military culture that made it unique: discipline, ethos, military customs, traditions and etiquette, esprit de corps. In his view, these elements of military culture were ways in which man made sense of war — they were a product of war, and in turn, also influenced "the likelihood and form of future wars". Commenting on the enduring nature of "military culture", Burk (1999) noted that even though the end of the Cold War brought a new emphasis by most military forces on missions such as peacemaking and peacekeeping, "warfighting still determines the central beliefs, values and complex symbolic formations that define military culture".

21ST CENTURY THREATS TO THE MILITARY PROFESSION

At this point, one should begin to appreciate that the concept of "the military profession" is a relatively recent product of human civilization in the last few centuries; and that this concept has meaning at both the macro, sociological or institutional level (i.e., as a certain form of organization of people in society) and at the psychological level (i.e., in terms of an individual's work or career-related experiences and attitudes). The latter idea that it is important to understand the military profession at both the macro and individual levels is echoed in a recent white paper published by the U.S. Army (2010) on the *Profession of Arms*, which aims to facilitate an organization-wide dialog on the subject. Figure 10-2 is an attempt to depict the relationship between individual members and their

FIGURE 10-2. Relationship between individual military professionals with the "profession of arms" (organization or institution) in each state.

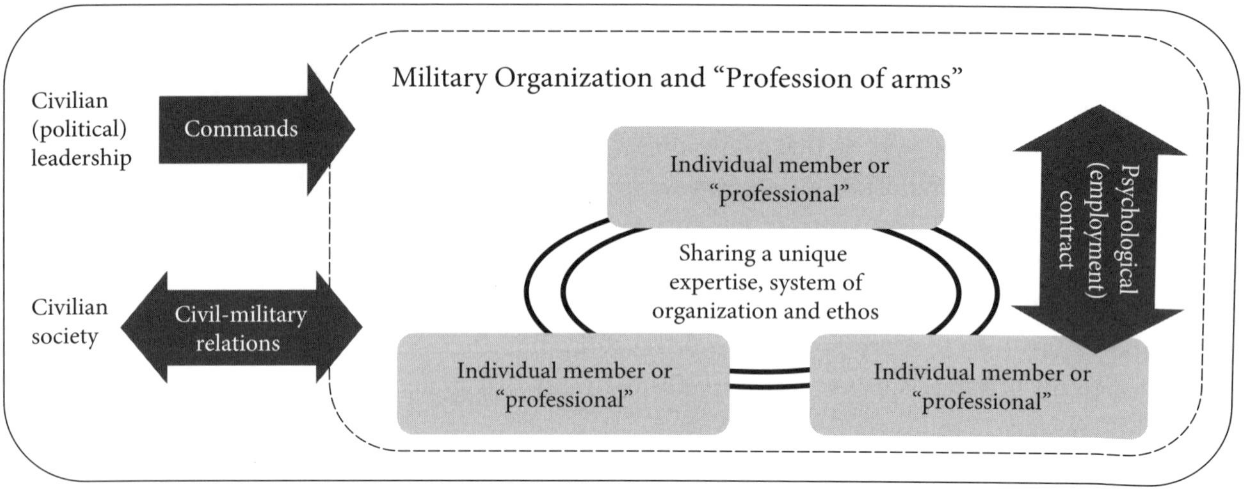

profession of arms — which is synonymous with the military organization in each state — and the nature of the relationship or "psychological contract" between the professional organization and the professionals, and the broader "relationship" with civil society in the state (i.e., civil-military relations). The next section discusses the 21st century "threats" to the military profession at both of these levels, as depicted in Figure 10-3. A common theme lies in the emergence of "boundarylessness" at the level of professional organizations or institutions, and, at individual-level professional careers.

Organizational-Level "Threats" to the Military Professional institution

Boundarylessness and the shift from Modern to Post-modern military organizational forms. In Chapter 9, we introduced the idea by organizational consultants, Ashkenas et al. (1995, 2001), who observed that since the 1980s, traditionally rigid, hierarchical organizations had shifted toward adopting more "boundaryless" characteristics and practices in order to achieve new success criteria like speed, innovation, flexibility and integration that were required for adaptiveness in a more dynamic business environment. This shift observed in general organizations was paralleled by observations by military sociologists — that with the end of the Cold War in the early 1990s, a new "post-modern" form of military organization was emerging in many developed and developing countries — one that was more "multipurpose" in mission, diverse in its personnel composition (i.e., with a greater mix of genders, ethnicity, uniformed and non-uniformed personnel, contractors, etc.), and where there was greater "permeability" with civilian society.

Writing on the post-modern military, Moskos, Williams, & Segal (2000) stated: "the dominant trend is a blurring of the lines between the military and civilian entities, both in structure and in

FIGURE 10-3. 21st century threats to the military profession.

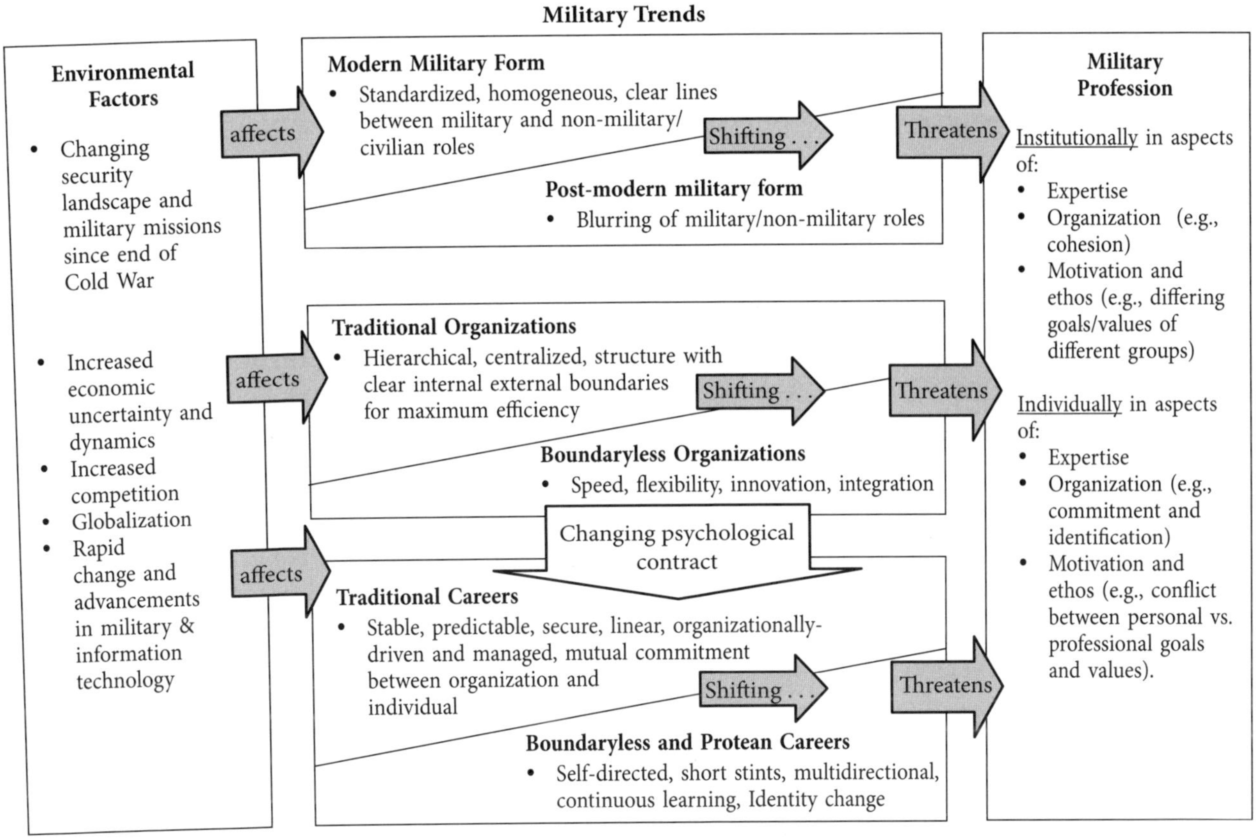

culture" (p. 11). They contrasted the "post-modern" military organizational form emerging in the 1990s with the "modern" military form that originated in the 19th century. Whereas the "modern" military forces generally maintained war-oriented structures and cultures that were distinct from civilian society, post-modern military forces were characterized by "increasing interpenetrability of civilian and military spheres, both structurally and culturally" (p. 2). Moskos et al. attributed the rise of the post-modern military form to cuts in military spending following the end of the Cold War when more states downsized their military forces and opted to operate in more multinational or coalition-type operations in peacekeeping roles as opposed to preparing primarily for conventional wars.

Individual-Level "Threats" to the Professional Military Career

Institutional-Occupational thesis by Charles Moskos. After the U.S. lost the war in Vietnam, sociologist Charles Moskos (1977) published a paper expressing concern that military service, especially among the U.S. military officers, seemed to be developing the characteristics of a civilian occupation as opposed to those of a profession. He was concerned that members of the U.S. military would come to see their service in the same way that any employee in a civilian organization would — where instead of being motivated by a commitment to serve society with ideals like "Duty, Honor, Country", that soldiers would become more concerned with pay, benefits, and the "quality of working life". He labeled these two perspectives the "institutional" versus the "occupational" models of work. In his view, it was society that formally recognized professions as societal institutions with norms, values and a purpose that required members of the profession to place larger societal interests above self-interests. In contrast, an occupational model of work was "legitimated" in terms of economic forces such as supply and demand, where self-interest and monetary rewards took priority over the larger interests of the society. If the institutional model focused on collective issues, the occupational model emphasized individualistic motivations rather than the good of the organization as a whole. Mosko's thesis was therefore especially related to the unique feature of the military as both a bureaucracy and a profession.

Subsequent research by Moskos (1986) showed that the internal attitudes in Western militaries were shifting from an institutional to an occupational framework. He was therefore concerned that military personnel might no longer be responsive to institutional practices, and cautioned that such creeping occupationalism would have negative effects on mission performance, soldier motivation, and the extent of professional responsibility. Moskos and Wood (1988) published an edited book which was provocatively titled *The military: More than just a job?* This book firmly established the Institutional-Occupational thesis as a framework for thinking about the threats to military professionalism at the individual level. Looking back over a decade later, Snider (1999) remarked in an article discussing the state of military culture in the U.S. forces, "Research conducted over the following decade with the I/O model helped the military make its transition to an all-volunteer force without eroding institutional norms and degrading overall professionalism" (p. 22). Subsequent research by Segal (1989) also suggested that soldiers could adopt both institutional and occupation orientations concurrently. Snider (1999) noted that in the post-Cold War era, military forces had to increasingly accept a more "pragmatic military professionalism" among career soldiers, balancing "concerns for both individual welfare and collective national security" (p. 22).

Rise of Boundaryless and Protean Career attitudes and the Changing Psychological Contract at Work. Just as Moskos' thesis about the post-modern military has some parallels with the general shift toward boundarylessness observed independently in organizations, his thesis on the institutional-occupational tension in military professionals can also be related to a general phenomenon also independently observed in the study of careers — the emergence of boundaryless, "Protean" careers and a general decline of traditional, organizational careers.

In 1976, Douglas Hall published *Careers in Organizations* where he speculated in his final chapter that a new form of career would emerge in the 21st century — one, which like Proteus, the Greek god of the Sea — could change its shape or form at will. He contrasted the protean career with the traditional, "organizational" career. In his view, the Protean career was one in which the person rather than the organization was "in charge", where workers would increasingly value "freedom" and "growth" at work, and, where the main criteria of career success were subjective (e.g., personal sense of accomplishment, work-life balance, etc.) rather than objective (e.g., advancement in organizational position and pay). The 1990s witnessed an increasing interest in D. Hall's idea of a "Protean career". With businesses experiencing a highly dynamic and uncertain environment in the 1990s, many organizations were forced to automate and downsize by laying off full-time employees and replacing them with part-time or temporary personnel who were typically offered limited job benefits and minimal job security. Rousseau (1989, 1995) noted a trend in the "new economy" — that there was a decreasing expectation on the part of both employers and employees for long-term, mutually satisfying relationships. Instead, both employers and employees showed increasing preference for short-term and highly transactional work arrangements.

In the 1990s, Arthur (1994; see also Arthur & Rousseau, 1996; Arthur, Inkson & Pringle, 1999) introduced the emerging concept of the "boundaryless career" — one with a strong preference organizational mobility and a positive attitude toward working across organizational boundaries. Research by Briscoe et al. (2006) showed that boundaryless career attitudes were related-to but independent of Protean attitudes in that the latter featured a somewhat individualistic preference for managing one's own career (as opposed to allowing the organization to manage one's career) and for making self-directed, personal values-driven career choices. As indicated in Figure 10-3, the shift toward more Protean and boundaryless career forms and mindsets has also been related to the idea of a weakening "psychological contract" between workers and their employers. Rousseau (1989, 1995) described "psychological contracts" as the implicit beliefs and expectations on the part of organizational employees of a certain "mutual obligation" between themselves and their employers. Psychological contracts were contrasted with what was explicitly written in formal employment contracts. There are two aspects of psychological contracts — a short-term or economic "transaction" aspect related to the exchange of employee effort and compensation by the employer, and a less tangible but long-term "relational" aspect that

The Chairman of an armed civilian contracting firm testifying before the U.S. Congress in 2007.

Source: Mark Wilson/Getty Images News/Getty Images

includes mutual trust, honor and "good faith". The creeping occupationalism as highlighted by Moskos in his Institutional-Occupational thesis and the general trend toward boundaryless and Protean careers in the world of work are both potentially individual level threats to the military profession in the 21st century — especially if they are not well-managed or influenced. While the former is a legacy of the tension between the bureaucratic nature of traditional military organizations versus the institutional ideals of the military profession, the latter is a general phenomenon reflecting global or worldwide changes in the nature of work and employment, of which soldiering is ultimately one of many other possible work or career options.

How rising "Boundarylessness" affects the 21st century Profession of Arms

Depending on the unique political, security and civil-military circumstances in each country, in the 21st century, the profession of arms and the individual members of the profession in each society have to cope with multiple "threats" that impact the professional expertise, institution or organization, and ethos or motivation while they focus on their primary military tasks and missions. The shift of military forces from the modern to the post-modern organizational forms, coupled with increasing demands for organizational "boundarylessness" and the rise of boundaryless, Protean careers in the general workforce is affecting the military profession in the following ways in the 21st century:

- **Threats to military professional expertise**: Unlike the modern military which had a clear focus on conventional warfare, post-modern military forces increasingly need to be prepared to operate in a broader range of operations including peacekeeping, humanitarian assistance, and low intensity conflict. The scope of professional military expertise has therefore widened to cater to this broader range of missions, and the boundary between traditionally military-only expertise and non-military expertise has been blurred. Unlike the "modern" military era where forces would strive to achieve self-sufficiency of expertise in its area of operations, "post-modern" military professionals need to work closely with other professionals such as the police, media, medical professionals, diplomatic corps, aid workers, etc., in order to achieve their mission. With smaller defence budgets, the military profession is less likely to be a life-long career — even for the full-time, "career" soldiers in most countries. Most military personnel may have to be prepared to take on a second career after their military service. Such employment arrangements may result in soldiers using more of their time in military service to prepare for a second career — which competes with their time required for honing their military expertise or for codifying or disseminating their military knowledge and skills to other professions. Budget cuts also drive military forces to hire more civilian contractors to release the uniformed personnel to focus on military training and operations (e.g., Cancian, 2008). The rapid increase in sophistication of military technology also demands more specialized expertise to maintain or operate certain military equipment — of which some of the expertise may only be available through non-military contractors. Together, factors like these can potentially threaten "expertise" as a basis for the military profession.

- **Threats to military professional organization**: The shift from modern to post-modern military forms brings along a greater diversity and heterogeneity of personnel which in turn makes cohesion-building within military units and formations a challenge. This challenge becomes even more critical when active and reserve military, military and non-military personnel (e.g., civilian contractors, medical professions, aid workers) are put together at short notice and expected to build "swift trust" during military operations (e.g., Ben-Shalom, Lehrer, & Ben-Ari, 2005). Forces that continue to take only a highly exclusive view of the military profession risk their military professionals becoming a minority in their organizations and operations, which in turn may weaken the overall respect and regard for military culture and traditions — especially among those who are "excluded" from the institution. The possibility of shorter military careers may also result in a rapid loss of professional institutional memory that is vital for helping new members to understand how the military works as a profession and why "things are the way they are". There is therefore a need to recognize the blurring organizational and career boundaries of the military profession in the 21st century, and to take a more inclusive approach to leading all members and affiliates of the profession while also clarifying and updating traditional military culture.
- **Threats to military ethos and motivation**: Driven by cuts in defence spending related to the ending of the Cold War in 1991, military forces have experienced a dramatic degree of downsizing since the late-1980s. Increasingly, it is common to find private contractors working alongside uniformed military personnel during military operations. Concerns have been expressed about the different motivations and governance of private contractors from uniformed military personnel — while contractors are motivated by money and profits and governed by private company regulations, uniformed military personnel are expected to be motivated by a sense of duty and are regulated by national laws (e.g., Schaub & Franke, 2009). A recent RAND survey found that the majority of lower ranking and younger U.S. military personnel familiar with the presence of armed private security contractors (PSCs) in their Iraqi operations felt that the higher pay of PSCs had a negative impact on their unit's morale, and, most of the soldiers believed that this would negatively impact military recruitment and retention (see Cotton et al., 2010). From a career management perspective, military forces increasingly find themselves having to compete in their own national workforce to attract and to retain talent. Many forces are resorting to the use of highly transactional but bureaucratically-efficient human resource tactics like offering enlistment or retention bonuses, and sponsoring non-military-related academic studies with the promise of post-military career employability, as a way to staff their profession. Such practices in turn have the potential to encourage more bureaucratic, occupational and managerial mindsets which in turn may threaten professional motivation especially among those who wish to see their military service as a professional calling.

STRATEGIES TO REVITALIZE AND STRENGTHEN THE PROFESSION OF ARMS

Before discussing the nature of leadership required in the 21st century profession of arms, it is useful to know that many nations and military forces have also started to take proactive measures to review, revitalize and refresh the profession of arms. In the case of some countries like Canada, professional military reform was in part driven by incidents that suggested a breakdown of professionalism in their forces (see Box 10-1). While some military forces have proactively introduced new doctrinal concepts and frameworks to deal with challenges facing the profession, and also enhanced the quality and system of professional education, countries such as the U.K. have also attempted at the national level to review and formalize aspects of the civil-military relationship. Each country and military forces' strategy for reforming its military profession is unique to its own circumstances. It is useful for military leaders to appreciate the many different strategies being adopted by nations or military forces to deal with 21st century challenges facing the military profession as they go about thinking about the nature of leadership and the roles that they need to play in the 21st century profession of arms (see Figure 10-4 for a summary of some of these strategies).

FIGURE 10-4. Strategies by military forces and nations to strengthen military professionalism.

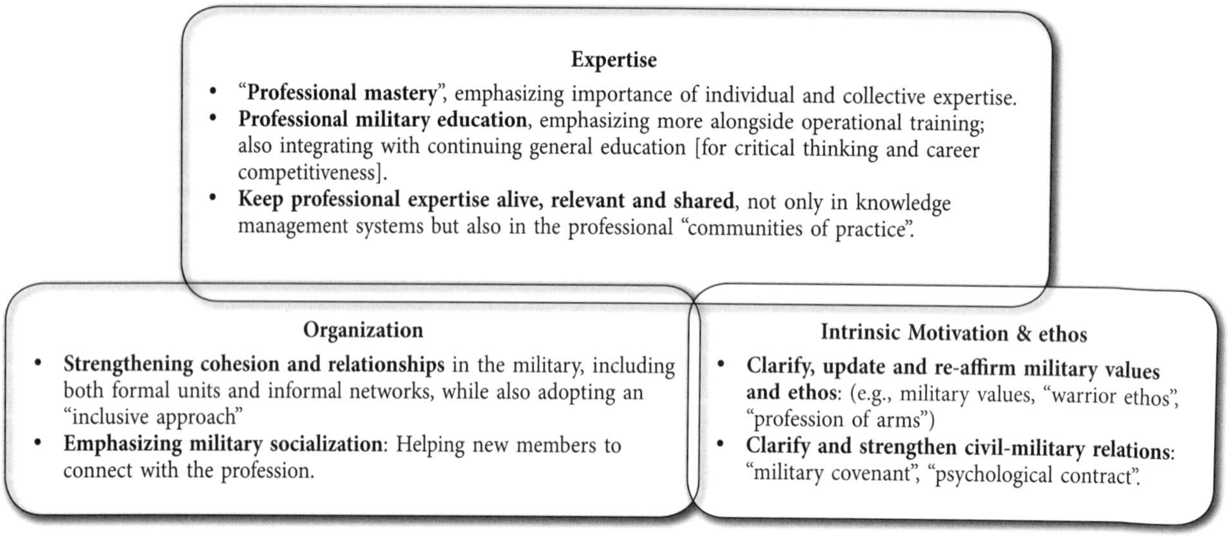

Strengthening Professional "Expertise"

Australian Defence Force's doctrinal emphasis on "Professional Mastery". Emphasizing the importance of both collective/organizational and individual-level professional expertise, the Australian

BOX 10-1

"Reprofessionalization" of Canadian Forces

In 1993, news broke out that Canadian soldiers sent on a peacekeeping mission in Somalia had beaten to death a 16-year old Somali youth who had broken into their compound. Writing in the Canadian Military Journal, Bercusson (1989) described how the four years following the media's revelation of the killing in 1993 led to the publication by Minister of National Defence of "Report to the Prime Minister on the Leadership and Management of the Canadian Forces" in 1997. This was "the darkest era in the history of the Canadian military in the post Second World War era". During this period, a Commission of Inquiry on the Somalia incident was established by the government, which "dissected" the Canadian military in public. Bercusson described how the incumbent Chief of the Defense Staff, an air force general, "squirmed on the witness stand and attempted to deflect the blame for failure to produce documents to his personal staff". Bercusson also remarked: "One of the main attributes of a profession is the ability to fairly and objectively govern itself. When the government appointed the Somalia Commission, it effectively suspended that condition of the profession of arms in Canada" (p.32). Today, the Canadian Forces has rebuilt itself as a profession. In Bercusson's (2009) analysis, two of the key interventions that helped the Canadian Forces come "up from the Ashes" were:

- **Professional Military Education reform.** This began with a revamp of the curriculum of the Royal Military College that made it compulsory for all military officers to take up courses in the arts, humanities and social sciences; the hiring of civilian academics with PhDs to teach in senior Canadian Forces programs and institutions; and the initiation of a professional (non-academic) Master of Defence Studies program.
- **Reform of professional military ethic and leadership.** The Canadian Forces Leadership Institute was established in 2002. One of its first efforts was to lead an organization-wide review and formulation of a new statement of the Canadian Forces ethos, called *Duty with Honor: The Profession of Arms in Canada* (see Canadian Forces, 2003). Its other significant effort was to develop a new Canadian Forces leadership doctrine to help guide continuous leadership development for its military personnel in the new operating environment (see Canadian Forces, 2005).

Defense Forces (2007) has introduced the concept of "professional mastery" in its joint doctrine, which states that "Professional mastery requires a certain critical mass in terms of experience, education and training whereby a group of individuals can exhibit a level of mastery in excess of the sum of the individuals' mastery". The Australian Defence Forces states that its emphasis on professional mastery is based on its "understanding of the chaos of war, and the recognition that morale and intellect are central to warfighting effectiveness", that "central to developing PM are its doctrine, education and training systems which "promote the individual's analytical and intuitive skills, create trust and build teams". The originators of the concept of professional mastery, de Somer and Schmidtchen (1999) noted that leaders at all levels will need to take a personal interest in guiding their people toward professional mastery, and in ensuring that people are selected for the right jobs. They must also understand the dynamics of professional mastery and the profound influence they have on the operational effectiveness of their units and organizations through the command climate that they create.

Holistic development of military professions: Complementing operational military training with professional military education and continuing general education. In the book *War and Anti-war* published just after the fall of the Berlin Wall, the futurists Alvin and Heidi Toffler (1993) wrote: "The new military needs soldiers who use their brains, can deal with a diversity of people and cultures, who can tolerate ambiguity, take initiative, and ask questions, even to the point of questioning authority." (p. 73–77). Increasingly, military forces have started to pay greater attention to complementing operational military training with professional military education. The U.S. military officially states that "professional development is the product of a learning continuum that comprises training, experience, education, and self-improvement", and that "professional military education provides the education needed to complement training, experience, and self-improvement to produce the most professionally competent (strategic-minded, critical-thinking) individual possible" (p. A-1; U.S. Chairman of the Joint Chiefs of Staff, 2009). The turn of the new millennium also witnessed various countries transforming their military training and education systems. Australia, Britain and Canada for example established large-scale Defense Academies that integrate previously disparate military training and education institutions, and introduced new military education and career management policies that together took a more holistic approach to the development of their military professionals by complementing operational military training with both professional military education and general academic education. These changes were driven by many factors — many of which were related to threats to the military profession described earlier in this chapter. These changes represent an increasing appreciation of the limitations of the traditional separation of military training from personnel management functions, and the need to take a more strategic approach to human capital development in the military profession.

Keeping professional military knowledge and expertise relevant, alive and shared in the community. Beyond merely emphasizing formal professional military education alongside operational military training, military forces are also connecting these with organizational learning systems to support knowledge creation, sharing and dissemination within professional military communities. Visible examples of such initiatives include the establishment of "lessons learned" centers close to military education and doctrine agencies, supporting "communities of practice" (e.g., U.S. Army's companycommand.com; see Dixon et al., 2005), establishing knowledge management systems to facilitate knowledge sharing and dissemination, and infusing learning capacities into military personnel and teams via practices like Senge's learning organization disciplines.

Strengthening Professional Ethos and Motivation

Re-emphasizing the professional military ethics. In recent years, military forces have launched various initiatives to emphasize the importance of professional military ethics to their troops, including efforts to introduce, update and discuss statements of professional military values, standards and creeds, and to enhance education on military ethics (e.g., Robinson, de Lee & Carrick, 2008. For example, the British Army in 2000 introduced a set of "Values and Standards" that made explicit previously assumed principles about how British military professionals were expected to behave (see Deakin, 2006; Mileham, 2008). These standards were revised and republished again in 2008 (see British Army, 2008). In 2003, the Canadian Forces published a document entitled *Duty with Honor: The Profession of Arms in Canada* that codified "for the first time" the attributed and values of the military profession. Stressing the importance of military ethics in its development of military leaders,

Members of the 1st Battalion, The Royal Anglian Regiment march through the highstreet during their homecoming parade in June 2010, after returning from their third tour of duty in Afghanistan.

Source: John Keeble/Getty Image News/Getty Images

the Australian Defence Forces, renamed its leadership centre located in its Defense College "Centre for Defence Leadership and Ethics" in 2009. The U.S. Army also adopted a "Soldier's Creed" and "Warrior Ethos" in January 2004. In 2007, it also appointed the U.S. Military Academy at West Point as the "Army Center of Excellence for the Professional Military Ethic", and entrusted it with the mission of assessing, studying, and helping to refine the professional military ethic of the Army, and, to accelerate moral ethical development among its individuals, units, and institutions (see Casey, 2009). In December 2010, the U.S. Army released a White Paper and Pamphlet to facilitate an organization-wide discussion on the nature of the army profession as part of a year-long Army-wide campaign to review this concept in an era of persistent conflict (see also Snider et al., 2009).

A national effort to strengthen civil-military relations: The British "Military Covenant". Besides formalizing its *Standards and Values* in 2000, the British Army also published a doctrinal document in 2000, entitled *Soldiering: The Military Covenant* which attempted to make more explicit the unique psychological contract between its soldiers, the military Service, and the Nation. The idea of the "Military Covenant" continues to be retained in the British Army's (2010) new *Operations* doctrine, the concept of the "Military Covenant" is also receiving the national level attention with the new British Prime Minister establishing a Task Force to look into "innovative ways" in which the different sectors of society could support the "Military Covenant". Some of the key recommendations of the Task Force include encouraging local communities and private companies to show greater public support and recognition toward military service members, veterans and their families, supporting service members' continuing academic education during service, and strengthening links between civilian and military sectors of society (see Strachan et al., 2010). While not explicitly focused on matters of professional military ethos and values, this national level effort helps to strengthen civil-military relations by raising national level awareness of the obligations of civil society to support the moral contract under which military personnel perform their professional duties.

Strengthening Professional Organization

Strengthening cohesion and professional networks. Research conducted during World War II and the Vietnam War (e.g., Shils & Janowitz, 1948; Henderson, 1985) demonstrated the importance of cohesion and esprit de corps in affecting military effectiveness in combat operations. While military cohesion refers to the feelings of identity and comradeship that is shared "horizontally" among soldiers in a unit or small group (i.e., horizontal cohesion), esprit de corps refers to the "vertical" or "upward" feelings of pride and commitment that the soldiers feel for their unit and larger formation (i.e., vertical cohesion). Burk (1999) regarded cohesion and esprit de corps as key elements of military culture; and, that they were affected by both structural arrangements in the organization (e.g., individual or unit rotation policies) and psychologically in terms values, beliefs, and emotions. Hence, military forces continue to emphasize the building and maintenance of cohesion and esprit de corps in their formal organizational structures such as units and formations. Increasingly however, military forces are also paying attention to supporting the professional cohesion emerging in informal networks — especially in the age of the Internet. A good example of this was the emergence of two websites — companycommand.com and platoonleader.org — which were first created informally by U.S. Army personnel in 2000 to facilitate peer-to-peer sharing of professional experiences and lessons learned from recent or current military operations (see Dixon et al., 2005). Their purpose was to help connect professional peers — past, present and future — by providing an informal forum to facilitate the sharing of tacit knowledge. Dixon et al., (2005) described how after two years of continuous growth as an informal network, the team that developed the network realized that it would need more organizational support to develop the network further. Today, these informal, professional forums are supported as part of the U.S. Army's internal knowledge management system.

Military socialization: Helping new members to connect with the profession. When individuals first join any organization, they typically experience anxieties such as "fitting in" to the new work culture, and where they may have to rethink some of their previous assumptions about "how things work". "Socialization" refers to the period and processes typically after selection and recruitment designed to help newcomers learn the social knowledge, skills, behaviors, values, beliefs, and attitudes so as to accept their new roles and function effectively within the organization (see McShane, 1998; Van Maanen, 1976). The learning involved in socialization can also be described as one form of organizational "sensemaking" where the newcomers begin to build a cognitive map of the organizational and job context, including their roles and responsibilities, their colleagues and partners, the organization including its history, mission, structure, politics, internal culture and external environment, the language, reward structures and rules. Research shows that newcomer learning results in better adjustment and job satisfaction. Individuals who are not able to socialize successfully are likely to leave the organization. With the typically high rate of recruitment and personnel turnover in military forces, and with threats to military professionalism due to increasing "boundarylessness", military forces are paying more attention to the socialization of their new members so as to prevent an erosion of the military's culture and ethos over time. Today, the Canadian Forces (2005) provides a definition of socialization in its leadership doctrine: "the formal and informal processes of teaching and persuading others to accept the core beliefs, values, behavioral norms, and roles of a particular culture". A U.S. Army Research Institute conducted a workshop on "Learning" which concluded that "Socialization happens whether one likes it or not", and, that "Leaders play a strong role in the process of socializing new organizational members (small unit, Big Army) and serve as a model of how to perform". The workshop called for military leaders to provide newcomers with "early, challenging assignments and mentoring lead to more effective organizational socialization" (see Quinkert, Morrison, Fletcher, Moses, & Roberts, 2007).

LEADING IN THE 21ST CENTURY PROFESSION OF ARMS: EMERGING THEMES

Chapters 8 and 9 highlighted the changing nature of military leadership in the complex, technical military operating environment and in context of traditionally hierarchical yet increasingly "learning" military organizations. This chapter has attempted to introduce another context for military leadership in the 21st century — that of the "military as a profession", while also highlighting certain "threats" and "challenges" facing the military profession and professionals in the new millennium, and various organizational and even national strategies to deal with these changes. What does all this mean for military leadership in the 21st century? Readers may notice the following themes emerging in this and the last two chapters:

Beyond leading people, military leaders must also "lead the professional institution". We believe that the Canadian Forces' (2005, 2007a, 2007b) articulation of "Leading People" and "Leading the Institution" in its leadership doctrine is one significant shift in the nature of military leadership required for the 21st century profession of arms. This framework shifts the focus of leadership thinking from one that is tied to levels in an organizational hierarchy to one that is more "distributed" throughout the organization and that serves different but related functions for the military. When leading the institution, the Canadian Forces (2007b) expects its military leaders to act as "stewards of the profession" by establishing and maintaining professional standards, and by strengthening and protecting professional standards, norms and values. The Canadian Forces (2007b) is very deliberate in stating the need for military leaders to "protect" the profession: "institutional leaders, in their role of steward, need to be constantly vigilant that military professional attitudes, norms and values (that is, professional ideology) remain pre-eminent when the ideologies of managerialism and entrepreneurialism, which are so influential in organizational governance, are confronted" (p. 5). In order to perform these roles, military leaders have to first possess a good understanding of military culture and ethos and the attributes of military professionalism. They must be able to articulate their profession's ideologies (i.e., values, ethos, creeds, etc.) not only to their soldiers but to external audiences, help to align their unit and organizational cultures with these ideologies, and embed these in the organization's culture.

Leadership must be a strong force for professional, organizational integration. Beyond building cohesion in formal military units and structures, 21st century military leaders must also play more facilitative roles to support the building of professional relationships and social networks. In an era of the more boundaryless professional military organizations and careers, military leaders need to play a more integrative role by constantly building a shared sense of vision, mission, values, purpose and ethos, and embedding throughout the organization the habits or "disciplines of the learning organization". Military leaders must also pay more attention to the socialization of all new military personnel. To do this effectively, they also need to actively learn about the military profession through professional military education programs. They also need to adopt a more inclusive view of the profession, and attempt to build relationships across different groups of personnel including active and reserve components, officers and non-officers, combat and non-combat/support vocations, civilian defense executives and uniformed personnel, in-service and retired personnel (and veterans), and "affiliates" of the military profession — especially military families. From a perspective of keeping professional knowledge alive and shared, military leaders will also need to actively participate in professional learning networks (or "communities of practice"). Writing on the nature of the leadership in a professional "complex adaptive system" (as opposed to a professional bureaucratic organization), Paparone et al., (2008) suggested eight new leadership tasks needed in such an environment. These included "relationship building, loose coupling, complicating, diversifying, sensemaking, learning, improvising, and emergent thinking" (p. 433), which is contrasted with the more "traditional" leadership tasks of "role defining, standardization, simplifying, socializing, decision making, knowing, commanding and controlling, and planning based on estimates" (p. 433).

Develop "meta-competencies" and engage in reflective practice for adaptive, professional leadership. In an era of ever changing missions and challenges, professions that fail to keep their core knowledge and expertise relevant and yet rigorous, current or alive yet shared within the community will fail in their mission and purpose. It is therefore important that military leaders themselves develop certain "meta-competencies" to enable them to lead adaptively in more complex and dynamic environments. According to Wong et al., (2003), Douglas Hall, who first introduced the idea of the "Protean Career" in 1976, played an influential role in the conclusions of a panel established by the U.S. Army to review the state of leadership development in its officer corps. Douglas Hall emphasized the importance of two "meta-competencies" which he believed were essential for enabling adaptive leadership (and for adapting to a more Protean work context): (1) identity, which includes one's self concept as a military professional, and (2) self-awareness, which includes an understanding of one's personal strengths and weaknesses. Today, the idea of "self-awareness" is featured in various military leadership doctrines including the U.S. and Australian doctrines (see Australian Defence Force, 2007; U.S. Army, 2006). The Singapore military's leadership framework explicitly features specific skills like "personal mastery", "self-awareness" and "self-management" as the "Self meta-competency" that develops other behavioral competencies (see Figure 1-2). From the perspective of constantly keeping one's professional/vocational knowledge and competency alive and relevant, another important leadership meta-competency could be "reflective practice" — a concept introduced by the organizational learning theorist Donald Schön (1983) in his book *The Reflective Practitioner: How professionals think in action*. Put simply, reflective practice refers to the habit of regularly and objectively reflecting or "looking back" at one's work — including the work processes and work situations — to consider how better to improve it. Reflective practices can include the habit of maintaining a personal journal to capture one's reflections, the practice of seeking feedback from others, Today, reflective practice is considered vital skill for continuous or life-long learning. Writers such a Paparone and Reed (2008) have suggested that military leaders also need to be reflective practitioners if they are to be effective members of a living and relevant profession of arms.

SUMMARY

While human civilization has always some form of organization of warriors to deal with basic security, "professions" and the "profession of arms" are relatively recent concepts. Just like other professions like medicine, law or engineering, the military is similarly characterized by a highly specialized knowledge base, a high degree of autonomy in evolving and applying that knowledge, and by the idea of service to society. Where the military profession is unique is the fact that the military profession and the formal military organization are synonymous in each state. To help explain why militaries "professionalized" and the unique relationship between the military and the state, Huntington (1957) suggested that professionalism is the key to civilian control over the military in modern democratic states. This is one of the dominant themes in the study of "civil-military relations" which is concerned with the relationship between the military services of a state with both its civilian (including political) authorities and with the larger civil society. Other unique aspects of the military profession are the concept of the soldier's "unlimited liability" (Hackett, 1983) and certain elements of military culture. This chapter has attempted to highlight three different clusters of "prototypical" features of professions — emphasizing that "military professionalism" goes beyond issues of technical expertise — it also concerns moral-ethical and organizational issues.

Modern military organizations constantly evolve their internal systems, structures and policies to adapt to new threats, missions demands, technologies and societal conditions. Military sociologists have studied how these internal and external changes often give rise to certain tensions in the military profession, and proposed some conceptual frameworks like the Institutional-Occupational thesis (Moskos, 1977, 1986), and the shift from the "Modern" to "Post-modern Military" (Moskos, Williams, & Segal, 2000) to facilitate better understanding of these military-specific trends. This chapter has also attempted to show how these military-specific tensions are also paralleled by more general trends toward "boundarylessness" in 21st century organizations (Ashkenas et al., 1995, 2002) and careers (Arthur, 1994). Together, these changes — both specific to the military and in general in the world of work and organizations — have the potential to threaten the military profession in terms of its expertise, moral-ethical and organizational aspects. It is evident from various strategies that military organizations and even national governments are concerned with the threat to their respective profession of arms. We note certain emerging themes related to the nature of leading in the 21st century profession of arms. These include the need for military leaders to "lead the institution", to act as a strong force for professional and organizational integration, and to develop certain "meta-competencies" like self-awareness and reflective practice in order that they lead more adaptively amidst the changing circumstances.

Key Concepts

- Profession and professionalism
- Professionalism
- Profession of Arms
- Modern/Post-modern military
- Civil-military relations
- Unlimited liability
- Boundaryless careers
- Protean careers
- Institutional-Occupational thesis
- Psychological Contract
- Military Socialization
- Professional military education
- Professional Mastery
- Psychological contract
- Military Covenant
- Military culture
- Metacompetencies
- Reflective Practice

Questions

- To what extent is the concept of the "military as a profession" a universal one that is commonly understood and applied across all countries? Are there certain universal norms (e.g., beliefs, values, traditions, practices, etc.) that make up the "military profession" in the world today? What would these be (if any)?
- To what extent has your military articulated the question of "who is/should be a military professional"? What is its position on whether non-combat officers (e.g., non-commission officers, military engineers, doctors and lawyers) are military professionals? Would it, for example, be important for a conscript force with large reserve components to have a clear position in this issue? Why?
- What is your military force's emphasis on military socialization and professional military education, especially among its personnel with leadership responsibilities? To what extent are these affected by the changing nature of the military in your society?

References

Arthur, M.B. (1994). The boundaryless career. *Journal of Organizational Behavior, 15(4)*, 7–22.

Arthur, M.B., Inkson, K., & Pringle, J. (1999). *The new career: Individual action and economic change*. London: Sage.

Arthur, M.B., & Rousseau, D.M. (1996). The boundaryless career as a new employment principle. In M.G. Arthur & D.M. Rousseau (Eds.), *The Boundaryless Career*, (pp. 3–20). New York: Oxford University Press.

Australian Defence Force (2007). *Joint operations for the 21st century.* Canberra: Strategic Policy Division, Department of Defence.

Ben-Shalom, U., Lehrer, Z., & Ben-Ari, E. (2005). Cohesion during military operations: A field study on combat units in the Al-Aqsa Intifada. *Armed Forces and Society, 32(1)*, 63–79.

Bercuson, D.J. (2009). Up from the ashes: The re-professionalization of the Canadian. Forces after the Somalia Affair. *Canadian Military Journal, 9(3)*, 31–39.

Briscoe, J.P., Hall, D.T., & DeMuth, R.L.F. (2006). Protean and boundaryless careers: An empirical exploration. *Journal of Vocational Behavior, 69*, 30–47.

British Army (2000). *Soldiering: The Military Covenant (Army Doctrine Publication No. 5).* London: Ministry of Defence (Army).

British Army (2008). *Values and Standards.* London: Ministry of Defence (Army).

British Army (2010). *Operations.* Shrivenham, Swindon, England: Development, Concepts and Doctrine Centre, British Ministry of Defence.

Burk, J. (1999). Military culture. In L. Kurtz & J.E. Turpin (Eds.), *Encyclopedia of Violence, Peace and Conflict.* San Diego, CA: Academic Press.

Canadian Government. (1997). *Dishonored legacy: The lessons of the Somalia Affair (Report of the Commission of Inquiry into the Deployment of Canadian Forces to Somalia).* Ottawa: Canadian Government Publishing.

Canadian National Defense (2003). *Duty with honor: The Profession of Arms in Canada (Summary).* Ottawa: Queen's Printer.

Canadian National Defense (2005a). *Leadership in the Canadian Forces: Doctrine.* Kingston, ON: Canadian Forces Leadership Institute, Canadian Defense Academy.

Canadian National Defense (2005b). *Leadership in the Canadian Forces: Conceptual Foundations.* Kingston, ON: Canadian Forces Leadership Institute, Canadian Defense Academy.

Canadian National Defense (2007a). *Leadership in the Canadian Forces: Leading People.* Kingston, ON: Canadian Forces Leadership Institute, Canadian Defense Academy.

Canadian National Defense (2007b). *Leadership in the Canadian Forces: Leading the Institution.* Kingston, ON: Canadian Forces Leadership Institute, Canadian Defense Academy. Available at: http://www.cda-acd.forces.gc.ca/cfli-ilfc/lea/index-eng.asp.

Cancian, M. (2008). Contractors: The new element of military force structure. *Parameters, 38(3)*, 61–77.

Carr-Saunders, A., and Wilson, P. (1933) *The Professions.* Oxford: Clarendon Press.

Casey, G.W. (2009). Advancing the Army Professional Military Ethic. *Joint Forces Quarterly, 54(3)*, 14–15.

Cotton, S.K., Petersohn, U., Dunigan, M., Burkhart, Q., Zander-Cotugno, M., O'Connell, E., & Webber, M. (2010). *Hired Guns: Views about armed contractors in Operation Iraqi Freedom.* Santa Monica, CA: RAND corporation.

de Somer, G., & Schmidtchen, D. J. (1999). *Professional Mastery: The human dimension of warfighting capability for the Army-After-Next* [Working Paper 107]. Canberra: Land Warfare Studies Centre.

Deakin, S. (2006). Ethics and the British Army's Values and Standards. *The British Army Review, 140*, 39–46.

Dixon, N., Allen, N., Burgess, T., Kilner, P., & Schweitzer, S. (2005). *Company Command: Unleashing the Power of the Army Profession.* West Point, NY: The Centre for the Advancement of Leader Development and Organizational Learning.

Eraut, M. (1994). *Developing professional knowledge and competence.* London: Falmer Press.

Flexner, A. (1915). Is social work a profession? *School and Society, 1*, 901–911.

Friedson, E. (2001). *Professionalization: The Third Logic.* Chicago: University of Chicago Press.

Gabriel, R., & Savage, P. 1978. *Crisis in Command.* New York: Hill and Wang.

Hackett, J. (1963). *The Profession of Arms.* London: Times.

Hall, D.T. (1976). *Careers in organizations.* Glenview, IL: Scott, Foresman.

Hall, D.T. (2002). *Careers in and out of organizations.* Thousand Oaks, CA: Sage.

Hall, R. (1968). Professionalization and bureaucratization. *American Sociological Review, 33*, 92–104.

Hart, B.H.L. (1953). *The Rommel Papers.* New York, NY: Harcourt, Brace and Company.

Henderson, W.D. (1985). *Cohesion: The human element in combat.* Washington, DC: National Defense University Press.

Holmes, R. (2005). *The soldier's trade in a changing world: Contract of unlimited liability.* Retrieved 17 June 2010 from http://www.bbc.co.uk/history/trail/wars_conflict/soldiers/soldier_trade_in_world_01.shtml.

Houle, C.O. (1981). *Continuing Learning in the Professions.* San Francisco, CA: Jossey-Bass.

Huntington, S.P. (1957). *The soldier and the state: The theory and politics of civil- military relations.* Cambridge, MA: Harvard University Press.

Kanter, R.M. (1989) Careers and the wealth of nations: a macro-perspective on the structure and implications of career forms. In M.B. Arthur, D.T. Hall and B.S. Lawrence (Eds) *Handbook of Career Theory.* Cambridge: Cambridge University Press.

Kerr, S., von Glinow, M.A., & Schriesheim, J. (1977). Issues in the study of "professionals" in organizations: The case of scientists and engineers. *Organizational Behaviour and Human Performance, 18*, 329–45.

MacArthur, D. (1962). *Farewell speech given to the Corps of Cadets at West Point on May 12, 1962.* Retrieved from http://www.nationalcenter.org/MacArthurFarewell.html.

MacDonald, K., & Ritzer, G. (1988). The sociology of the professions: Dead or alive? *Work and Occupations, 15(3)*, 251–272.

McShane, S.L. (1998). *Canadian organizational behaviour. (3rd ed).* Toronto: McGraw-Hill Ryerson.

Mileham, P. (2008). Teaching military ethics in the British Armed Forces. In P. Robinson, N. de Lee, & D. Carrick (Eds.), *Ethics education in the military.* Aldershot, England: Ashgate.

Mileham, P. (2010). Unlimited Liability and the Military Covenant. *Journal of Military Ethics, 9(1)*, 23–40.

Millerson, G. (1964). *The qualifying associations.* London: Routledge & Kegan Paul.

Moskos, C.C. (1977). From institution to occupation: Trends in military organization. *Armed Forces & Society, 4(1)*, 41–50.

Moskos, C.C. (1986). Institutional/Occupational trends in armed forces: An update. *Armed Forces & Society, 12(3)*, 377–382.

Moskos, C.C. (1988). *Soldiers and sociology.* Evanston, IL: US Government Printing Office.

Moskos, C.C., Williams, J.A., & Segal, D.R. (2000). *The Postmodern military: Armed Forces after the Cold War.* NY: Oxford University Press.

Moskos, C.C., Wood, F.R. (Eds., 1988). *The military: More than just a job?* Washington, DC: Pergamon-Brassey's.

Paparone, C.R., Anderson, R.A., & McDaniel, R.R. (2008). Where military professionalism meets complexity science. *Armed Forces & Society 34(3)*, 433–449.

Paparone, C., & Reed, G. (2008). The Reflective Military Practitioner: How Military Professionals Think in Action. *Military Review, 88(2)*, 66–75.

Quinkert, K.A., Morrison, J.E., Fletcher, J.D., Moses, F.L., & Roberts, E.J. (2007). *The Army Science of Learning Workshop* (ARI Research Note 2007-02). Arlington, VA: United States Army Research Institute for the Behavioral and Social Sciences.

Robinson, P., de Lee, N., & Carrick, D. (2008). *Ethics education in the military*. Aldershot, England: Ashgate.

Rousseau, D.M. (1989). Psychological and implied contracts in organizations. *Employee Rights & Responsibilities Journal, 2*, 121–139.

Rousseau, D.M. (1995). *Psychological contracts in organizations; Understanding written & unwritten agreements.* Thousand Oaks: Sage Publications.

Sanders, C.R., & Lyon, E. (1976). The humanistic professional: The reorientation of artistic production," pp. 43–59 in J.E. Gerstl & G. Jacob (Eds.) *Professions for the People: The Politics of Skill.* Cambridge, MA: Schenkman.

Schaub, Jr., G., & Franke, V. (2009). Contractors as Military Professionals? *Parameters, 39(4)*, 88–104.

Schon, D.A. (1983). *The Reflective Practitioner: how professionals think in action*. London: Temple Smith.

Schon, D.A. (1987). *Educating the reflective practitioner.* San Francisco, CA: Jossey Bass.

Segal, D.R. 1989. *Recruiting for Uncle Sam: Citizenship and Military Manpower Policy*. Lawrence: University Press of Kansas.

Shils, E., & Janowitz, M. (1948). Cohesion and disintegration in the Wermacht in World War II. *Public Opinion Quarterly, 12*, 280–315.

Snider, D.M. (1999). An uninformed debate on military culture. *Orbis, 43*, 11–26.

Snider, D.M., & Matthews, L.J. (eds., 2002). *The Future of the Army Profession* (1st Ed.). NY: Mc-Graw-Hill.

Snider, D.M. (2005). The army as profession. In D.M. Snider, D.M. & L.J. Matthews (Eds.), *The Future of the Army Profession (2nd ed.).* NY: McGraw-Hill.

Snider, D.M., Oh, P., & Toner, K. (2009). *The Army's Professional Military Ethic in an era of persistent conflict.* Carlisle, PA: Strategic Studies Institute, U.S. Army War College.

Strachan, H., Armour, T., Healy, P., & Smith, M. (2010). *Report of the Task Force on the Military Covenant.* London: U.K. Ministry of Defense.

Toffler., A. & Toffler, H. (1993). *War and anti-War.* Boston, Mass.: Little, Brown, and Co.

U.S. Army.(2001). *The Army Training and Leader Development Panel Officer Study Report to the Army.* Washington, DC: Headquarters, Department of the Army.

U.S. Army. (2005). *The Army (Field Manual 1).* Washington, DC: Headquarters, Department of the Army.

U.S. Army. (2010). *The Profession of Arms: An Army white paper.* Washington, DC: Headquarters, Department of the Army.

U.S. Chairman of Joint Chiefs of Staff (2009). *Officer Professional Military Education Policy (OPMEP; Instruction 1800.01D).*

Van Maanen, J. (1976). Breaking in: Socialization to work. In R. Dubin (Ed.), *Handbook of work, organization, and society*, p. 67. Chicago: Rand McNally.

Vollmer, H.M., & Mills, D.L. (eds., 1966). *Professionalization*. Englewood Cliffs, NJ: Prentice-Hall.

Wong, L., Gerras, S., Kidd, W., Pricone, R., & Swengros, R. (2003). *Strategic leadership competencies.* Carlisle, PA: Strategic Studies Institute, US Army War College.

Index

A

AARs. *see* After-action reviews
Abilities, 97
  and competencies, 64
Adaptive expertise, 173–174
Adaptive team model, 181
Adaptive team performance, 173
  leader actions for, 187
Adaptive teams, 175
Administrative Science (Fayol), 199
After-action reviews (AARs), 212, 216
All We Could Be: The Rebuilding of the Army from Vietnam to Desert Storm (Brown), 213
American Soldier (Stouffer), 16, 19, 140
Approach causes, 72
Army Alpha test, 106
Army Beta test, 106
Assumptions, 64–65
Attitudes, 66
Australian Defence Force, 3, 8, 129, 132, 209, 238–240
  on command, leadership and management, 13–14
Australian Defence Force Psychology Organization, 34
Authentic leadership, 132–134
Autocratic leaders, 127
Avoidant causes, 72

B

Bass' model of transformational leadership, 122–124
Bass' theory, military research, 124–126
Battlemind, 40–41
Behavior, 55
  human psychology, 54–55
  of soldiers, 61, 62
Behavioral approaches, 6
  early trait and, 94–96
Behavioral science, 43–44
Beliefs, 64–65
BICEPS. *see* Brevity, Immediacy, Contact, Expectancy, Proximity & Simplicity
Big Five personality factors, 98, 99
Boundaryless careers, 235
Boundarylessness, 21st century profession of arms, 236–237
Boundaryless organization, 207
Brevity, Immediacy, Contact, Expectancy, Proximity & Simplicity (BICEPS), 43
Bridge resource management (BRM), 172
Built to Last (Collins), 200
Burns on transforming leadership, 121
Bystander effect, 60, 62

C

Canadian Forces (CF), 3, 13–14, 128–129
  on command, leadership and management, 13–14
  definition of leadership by, 206, 216
  leadership levels in, 206
  reprofessionalization, 239
Can do traits, 97, 98
Careers in Organizations (Hall), 234
Casualties
CSR and, 38–39
destruction and, 35–36
Category-based trust, 152
CF. *see* Canadian Forces
Character perspective on values, 131
Civilian involvement, 31–32

Civil-military relations, 16, 228
  national effort to strengthen, 242
Classical decision making (CDM), 176
Coach, 188
Cognitive ability, 97
Cognitive complexities, 168
Cognitive Resource theory of leadership, 100
Cognitive skills for military leadership, 103–104
Cognitive task analysis (CTA), 177
Cohesion, 80–81
  strengthening, 242–243
Collective leadership, 207
Combat motivation, 79–84
Combat Stress Reaction (CSR), 38–39
  management, 41–43
Command
  responsibility, 39–40
  vs. leadership, 12–14
Command information center (CIC), 168
Command Team Effectiveness Framework (CTEF), 146–147
Competencies, abilities and, 64
Competency modeling, 104–105
Complex security threats, 28–30
Conscientious objectors, 83
Consideration, leadership behavior, 95–96
Context-free knowledge *vs.* context-specific knowledge, 9–10
Context-specific knowledge *vs.* context-free knowledge, 9–10
Contextual performance, 51–52
Contingency theory, 96, 119
Continuum of styles, 117–118
Counterproductive Work Behavior (CWB), 52–53
Crew Resource Management (CRM), 141, 172
Criterion of leadership, 7–8
Critical thinking skills, 173, 176–178
CRM. *see* Crew Resource Management
Cross training, 186–187
CSR. *see* Combat Stress Reaction
CWB. *see* Counterproductive Work Behavior

D
Decision-making theories, 176–177
Defence Science and Technology Organization, 34
Deindividuation effect, 60
Democratic leaders, 127
Deprivation, 34–35
Destruction, 35–36
Devil's Advocate Technique, 191
Diffusion of responsibility, 62
Direct-level leadership, 203–204, 206
Disposable leadership, 207
Distributed leadership, 206, 214
Doctrinal *vs.* academic definitions, 8–9

E
Effective leadership, 216
Effort-performance (E-P), 76
Emerging technologies, 31
Enemy threat, 35
E-P. *see* Effort-performance
Equity theory, 77–78
Esprit de corps, 84
Ethical reasoning perspective on values, 131–132
Expectancy theory, 76–77
External adaptation, 175

F
Facilitator, 190
Factor analysis, 95
Factors influencing morale, 87, 89
Fatigue, 34–35
Fear, 37, 45
Fiedler's contingency theory of leadership, 96, 119
Five Factor Model (FFM) of personality, 98, 99
Fog of war, 35
Fragging, 82
Full Range of Leadership training programs, 124–126, 134
Functional leadership theory, 149–150
Fundamental attribution error, 6, 63

G
Good to Great (Collins), 200
Great Man theory, 101, 110
Grid of styles. *see* Continuum of styles
Group decision support systems (GDSS), 168
Group informational influence, 58–59
Group/unit normative influence, 59
Guided Team Self-Correction Techniques, 182–183

H
Handbook of Leadership, (Bass), 5

Handbook of Leadership (Stogdill), 94
Hardiness, 40–41
Harsh physical environment and working conditions, 33–34
Herzberg two-factor theory, 74–75
Hitler's leadership, 66, 133
Human behavior
  causes of, 72
  psychology of, 54–55
Human in Command, 12–13
Human in Command: Exploring the Modern Military Experience (McCann & Pigeau), 17
Huntington's thesis, 227–228

I
Ideology, 83–84
IDF. *see* Israeli Defense Force
Ill-defined enemy, 30
Images of Organization (Morgan), 198
Implicit communication, 179
Indirect leadership, 206
Individual differences, 97, 99
Industrial-organisational psychology, 54
Information Provision phases, 185
Initiating structure, 95, 96
Input-Process-Output (I-P-O) model, 143–144
Institutional-occupational thesis, 233–234
Instructor, 188
Integrative approach, 7
Internal adaptation, 175
Internal control among soldiers, 130
Interpositional knowledge (IPK), 186
Irregular warfare (IW), 28–29
Israeli Defense Force (IDF), 87, 125
  organizational learning, 212
IW. *see* Irregular warfare

K
Knowledge, 63–64
Knowledge, skills and abilities (KSA), 148

L
Laissez-faire leadership, 123
Leader actions
  critical thinking, 190–191
  develop adaptive expertise, 188–190
  SMM, 191
  team learning, 188
  team self-diagnosis, 192
Leader attribution error, 6
Leader behaviors, 150–151
Leader-led after-action debriefing (LLAAD), 43–44
Leadership, 81–82
  in academic textbooks, 4
  direct-level, 203–204, 206
  disposable, 207
  distributed, 206, 214
  effectiveness, 8, 216
  fundamental attribution error, 6
  Handbook of Leadership, (Bass), 5
  Handbook of Leadership (Stogdill), 94
  Hitler, 66, 133
  leader attribution error, 6
  military and non-military definitions, 2–3, 5
  in military doctrines, 3
  power, 7
  Romance of Leadership, 6
  standard/criterion of, 7–8
  vs. command, 12–14
  vs. management, 10–11
Leadership behaviors, early search for, 95–96
Leadership development, military, 108–110
Leadership 24–7 framework, 8–9
Leadership potential, 106, 108
Leadership qualities, early search for, 94–95
Leadership styles, 116–120, 147
  Managerial Grid theory of, 118
  military forces on, 127–129, 135
  new paradigm, 120–126
  situational factors, 126
  and values, 132–134
Learning organization
  cultural norms of, 213
  demystifying, 212–213
  disciplines of, 209–210
  U.S. army's transformation, case study, 213–215
Levels of leadership, 203, 205, 206
Life and death nature, 130
LLAAD. *see* Leader-led after-action debriefing

M
Management, 7
  vs. leadership, 10–11
Managerial Grid theory of leadership styles, 118

Maslow's hierarchy of needs theory, 73–74
basic biological and physiological needs, 80
Maximum performance, 51
Mental health professionals, 44
Mentor, 188
Meta-analysis, 96, 98, 100, 124
Meta-competencies, 105, 245–246
Metarecognition, 174, 177
Military
as bureaucratic/professional institution, 230–231
ethos/motivation, threats to, 237
forces, professionalize, 227–228
as profession, 227–231
Military covenant, 242
Military culture, 231
Military doctrines, 3
Military ethics, 132
Military ethos, 16
Military forces on leadership styles, 127–129, 135, 215–216
Military leaders
levels of, 202–206
role of, 215
selection, 105–107, 110
Military morale, 15
Military Operations Other Than War (MOOTW), 29
Military organization
forms of, 206–208
learning, 208–209
Military profession, 228–229
holistic development of, 240
21st century threats to, 231–233
unique features of, 229–231
Military professional expertise
threats to, 236–237
Military professional institution. organizational-level threats, 232–233
Military professional organization, threats, 237
Military professionals *vs.* profession of arms, 232
Military socialization, 243
Modern/post-modern military, 232–233
MOOTW. *see* Military Operations Other Than War
Morale, 85–89
Motivation, 15, 67
combat motivation, 79–84
definition, 72
equity theory, 77–78
expectancy theory, 76–77
Herzberg two-factor theory, 74–75
Maslow's hierarchy of needs theory, 73–74
morale, 85–89
nature of, 72
reinforcement theory, 75
self-determination theory, 78–79
Motivational-developmental theory, 100–101
Motivation-hygiene theory, 74–75
Motivation to lead (MTL), 100
Multifactor Leadership Questionnaire (MLQ), 122
Multiple security threats, 28–30

N
Naturalistic decision making (NDM), 176–178
Nature of leadership in organizations, 200–202
Nature of Organizational Leadership (Zaccaro), 200
Negative task performance, 51
New Leadership approach, 7
Non-ability traits, 97
Non-leadership. *see* Laissez-faire leadership

O
Office of Naval Research (ONR), 141
Ohio State Leadership Studies (Stogdill), 95
OLMs. *see* Organizational learning mechanisms
Operation Iraqi freedom
Abu Ghraib prisoners abuse, 61
U.S. soldiers in, 80, 83
mental health of, 39
Organization
definition of, 198
nature of leadership in, 200–202
principles of, 199
Organizational climate, 200
Organizational culture, 199–200
definition of, 199
elements of, 199–200
Organizational Culture and Leadership (Schein), 200
Organizational development (OD), 140, 153, 210–211
definition of, 211
Organizational effectiveness (OE), 153
Organizational leaders
activities of, 201
capabilities of, 201
challenges of, 202

Organizational learning, 210
aspects of, 212
concept of, 211
single and double-loop learning, 211
Organizational learning mechanisms (OLMs), 212, 213
Organizational-level leadership, 204
Organizational structure, 198–199
Organizational Theory (Robbins), 198

P
Pattern recognition, 177
Peer ratings technique, 20
Performance, classification of, 51
Performance-outcome (P-O), 76
Personality, 66–67, 97–99
Person based trust, 152
Person-focused leadership behaviors, 151
Peter Senge's "five disciplines," 209–210
PIE. *see* Proximity, Immediacy, Expectancy
P-O. *see* Performance-outcome
Positional clarification, 186
Positional modeling, 186
Positional rotation, 186–187
Power, 7
Principles of Scientific Management (Taylor), 10, 198
Private security contractors (PSCs), 237
Process gains, 144
Process loss, 144
Profession
beliefs of, 225
characteristics of, 224
concept of, 222
definition of, 222–223
developmental perspective, 225
as distinct type of work, 223–224
as dynamic process, 225
history of, 223
military as, 227
prototypical features of, 226–227
Professional ethos/motivation, strengthening, 241–242
Professional expertise, 226, 238–240
strengthening, 238–240
Professionalism, 222
definition of, 223
psychology of, 225–226
Professionalization, as on-going process, 225
Professional mastery, 238–240
Professional military career, individual-level threats, 233–236
Professional military education, 240
reform, 239
Professional military ethic/leadership, reform of, 239
Professional military ethics, re-emphasizing, 241–242
Professional motivation/ethos, 227
Professional networks, strengthening, 242–243
Professional organization, 227
strengthening, 242–243
Profession of arms
21st century, 244–246
strategies to revitalize and strengthen, 238–243
vs. military professionals, 232
Protean careers, 234–236
Proximity, Immediacy, Expectancy (PIE), 43
PSCs. *see* Private security contractors
Pseudo-transformational leadership, 133
Psychological contract, 234–236
Psychological operations, 83
Psychology of human behavior, 54–55

R
Reflective practice, 245–246
Reinforcement theory, 75
Resilience, 39–44
Role influence, 59–60
Romance of Leadership, 6

S
SAF. *see* Singapore Armed Forces
Scientific approach, 21
Scientific knowledge, 19
Selection
military leader, 105–107
vs. classification, 94
Self-confidence, 82–83
Self-determination theory, 78–79
Self-monitoring trait, 96
self-regulatory skills, 174
Shared mental models (SMM), 148, 173, 179–180
Singapore Armed Forces (SAF), 3, 129, 216
Single *vs.* double-loop learning, 211
Situational/contingency approach, 6
Situational Leadership theory, 118–119

Skills, 63–64
acquisition phases, 185
Socialization, military, 243
Social-psychological perspectives, 140
Social sciences, 19
Social-scientific approach, 19–20
Soldiers' performance
contextual performance, 51–52
CWB, 52–53
framework of, 53–54
situational factors
deindividuation effect, 60
fundamental attribution error, 63
group informational influence, 58–59
group/unit normative influence, 59
operation Iraqi freedom, Abu Ghraib prisoners, 61
role influence, 59–60
Vietnam War (1968), 56–57
weak and strong situations, 62–63
task performance, 50–51
Story-Test-Evaluate-Plan (STEP) process, 178
Strategic-level leadership, 204
Stratified systems theory, leadership levels, 205
Strength-Weaknesses-Opportunities-Threat (SWOT), 156
Stress
definition of, 37
management and resilience
behavioral science and mental health professionals, 44
command responsibility, 39–40
CSRs, 41–43
hardiness and battlemind, 40–41
traumatic event management, 43–44
model of, 36
process, 37
response, 37–38
Stress exposure training (SET), 183–185
phases of, 185
Stress management techniques, 40
Stressors, 30, 33–36, 39
Stress outcomes, 37
Styles approach, origins of, 116
Surface Warfare Officers School (SWOS), 191
Swift trust, 151
Systematic team building, 156

T

Tactical Decision Making Under Stress (TADMUS), 168
research program, 141
Task-focused leadership behaviors, 151
Task performance, 50–51
Task, Relationship and Change (TRC) behaviors, 102–103
Task work dimension, 154
Task work processes, 144
Team adaptation and coordination training (TACT), 181–182
Team building
effectiveness of, 155
personality tools in, 157
systematic team building, 156–157
technique of, 154–155
Team decision making, critical thinking skills for, 176–178
Team Dimensional Training (TDT), 182–183
Team effectiveness models
I-P-O model of, 143
leadership factor in, 147–148
Team leaders, implications for, 170–171
Team research, evolution of, 140–142
Teams
coordination, 174
development, 157–158
efficacy, 174
groups *vs.*, 142
interventions, 145–146
leader behaviors in, 150–151
leadership, 147, 148
outcomes, 154
processes, 145–146
self-diagnosis, 192
skills, 144–145
teamwork, 142–144
trust in, 152–153
Team training, 171–173
strategies for, 180
Teamwork, 142–144
dimension, 154
Tele-leadership, 207
The Boundaryless Organization: Breaking the Chains of Organizational Structure (Ashkenas), 206

The Death and Birth of O.D. (Tichy), 214
The military: More than just a job? (Moskos), 234
The Profession of Arms (Hackett), 229
The Reflective Practitioner: How professionals think in action (Schön), 245
The Soldier and the State (Huntington), 227–229
Three Block War, 29
Timespan of responsibilities, 205
Training critical thinking skills, 177–178
Trait approach, 6
  to professions, 223–224
Traits, 100
  early, 94–96
  enhanced understanding of, 97–99
  and military leader selection, 105–107
  relating to leadership, 100–101
Transactional leadership
  affecting unit performance, 124
  contingent reward, 122
  management by exception-active, 122
  management by exception-passive, 122
Transformational leadership
  affecting unit performance, 124
  bass' model of, 122–124
  burns on, 121
  idealized influence, 122
  individualized consideration, 122
  inspirational motivation, 122
  intellectual stimulation, 122
  from transactional to, 120
Traumatic event management, 43–44
Tuckman's stage theory, 157–160
Typical performance, 51

U
U.K. Defence, 3
Unlimited liability, 229–230
U.S. Army, 3
  on improving organization, 215–216
  leadership doctrine, levels, 203–204
  levels of leadership, 205
  OE model, 214
  post-Vietnam organization development initiatives, 214
  transformation to learning organization, case study, 213–215
U.S. Army doctrine, styles in, 127–128
USS Vincennes, 168, 169

V
Values, 64–65
  and ethics, 131–132
  and military leadership, 129–132
Values-based leadership, 132, 135
Vietnam War (1968), My Lai massacre of civilians, 56–57
Vroom's expectancy theory, 76

W
War and Anti-war (Wall), 240
Weak and strong situations, 62–63
Will do traits, 97, 98